THE PHYSIOLOGY
OF PLANTS UNDER
STRESS

WITHDRAWN

THE PHYSIOLOGY OF PLANTS UNDER STRESS

Maynard G. Hale
Emeritus Associate Professor of Plant Physiology

David M. Orcutt
Associate Professor of Plant Physiology

Department of Plant Pathology, Physiology, and Weed Science
Virginia Polytechnic Institute and State University
College of Agriculture and Life Sciences
Blacksburg, Virginia

with a chapter by
Laura K. Thompson on Irradiation Stress

A WILEY-INTERSCIENCE PUBLICATION
JOHN WILEY & SONS
New York • Chichester • Brisbane • Toronto • Singapore

Library of Congress Cataloging-in-Publication Data:

Hale, Maynard G.
 The physiology of plants under stress / Maynard G. Hale and David
 M. Orcutt; with a chapter by Laura K. Thompson on irradiation
 stress.
      p. cm.
   Includes bibliographies and index.
   ISBN 0-471-88997-0.   ISBN 0-471-63247-3 (pbk.).
 1. Plant physiology. 2. Plants, Effect of stress on. I. Orcutt,
 David M. II. Thompson, Laura K. III. Title.
 QK711.2.H35 1987
 581.2′4—dc19   87-17609
                                                     CIP

Printed in the United States of America

10 9 8 7 6 5 4 3

PREFACE

As the human population of the world increases and utilizes more land area for housing and industrial activities, agriculture is being forced into marginally productive areas. Drought, nutrient deficiencies and toxicities, salinity, temperature extremes, air pollution, and chemical interference are stresses often encountered. Alteration of environments and climates may also result from human activities that increase the stressful conditions under which plants must grow and survive.

Knowledge of the effects of various stresses on the physiology of plants is essential to an understanding of resistance and survival mechanisms and to breeding for stress resistance. Selection by humankind is more rapid for crop plants than the natural selection of evolution. Development of new cultural practices using technology to alleviate stress effects also depends upon a knowledge of the physiological reactions of plants to stressful conditions. Plants have characteristics that may enable them to survive aberrant metabolism, hormonal imbalances, and membrane disfunctions. The processes of tolerance and avoidance of the effects of stress are not completely known but there is a large body of knowledge accumulating, which we have attempted to summarize for those students in plant science who have a basic understanding of plant physiology.

For a number of years, the senior author has successfully taught a graduate level course in plant stress physiology, and the authors collectively have many years of experience in organizing and teaching courses in various aspects of plant physiology for all levels of students in agricultural curricula. This experience has been useful in writing with the student in mind.

The questions at the ends of chapters should be useful in stimulating discussion and thought. General and specific references have been carefully chosen from the vast literature for their clarity and pertinence to the state of knowledge.

We acknowledge the inspiration of all those students who have passed through our courses and who left something of themselves with us. It has been their enthusiasm and quest for knowledge that has encouraged us to attempt this textbook. We also thank our colleagues who have continued to encourage us and critically listen to our expositions of stress concepts and principles.

MAYNARD G. HALE
DAVID M. ORCUTT

Blacksburg, Virginia
September 1987

CONTENTS

1 TERMINOLOGY   **1**

References   4

2 DROUGHT STRESS   **5**

Basic Water Relations Concepts   5
Effects of Drought on Growth and Yield   11
Effects on Ultrastructure   15
Effects of Water Stress on Photosynthesis   16
Nitrogen Metabolism under Water Stress   20
Water Logging and Anaerobiosis   21
Discussion Questions   22
References   23

3 DROUGHT RESISTANCE   27

Osmotic Adjustment   27
Cell Elasticity   33
Drought Escape   34
Drought Tolerance   35
Measurement of Drought Stress   37
Breeding for Drought Stress Resistance   39
Discussion Questions   41
References   42

4 TEMPERATURE STRESS   45

Chilling Injury   46
Effects on Membranes   47
The Freezing Process   47
Freezing Injury to Membranes   49
Effects of Freeze–Thaw Cycles on Plasma
Membranes   50
Tolerance of Freezing Stress   52
Frost Resistance and Cold Hardiness   54
Effect of Temperature on Root Processes   55
Breeding for Temperature Tolerance   57
Effects of High Temperature   58
Temperature Acclimation   60
Acclimation of Photosynthesis   61
High-Temperature Acclimation   63
Acclimation to Low Temperature   64
Factors Affecting Cold Hardening   65
Discussion Questions   66
References   67

5 NUTRIENT STRESS   71

Conditions Causing Nutrient Stress   72

Deficiency Causes and Symptoms   75
Deficiency and Toxicity Causes and Effects   77

    Copper, 77
    Manganese, 78
    Iron, 78

Plant Analysis as a Diagnostic Tool   79
Nutrient and Metal Toxicity   83
Aluminum Toxicity   84
Manganese Toxicity   85
Copper Toxicity   86
Mycorrhizae as a Factor in Stress Alleviation   87
Chelation as a Mechanism of Tolerance   87
Genetics of Mineral Element Stress Tolerance   88
Discussion Questions   89
References   90

6 SALT STRESS   **93**

Mechanisms of Tolerance   95
Breeding for Salt Tolerance   97
Specific Ion Effects   99
Discussion Questions   99
References   100

7 IRRADIATION STRESS   **103**

Atmospheric Attenuation of Solar Radiation   103
Distribution of Radiation in a Plant Community   104
Uptake of Radiation by Plants   106
Sun versus Shade Plants   106
Effects of Light Deficit (Shade)   109
Effects of Bright Light   110
Resistance to High Light Intensity Injury   110
Ultraviolet Radiation   111

Ionizing Radiation as Stress   112
Discussion Questions   113
References   114

8 ALLELOCHEMICAL STRESS   117

The Juglone Story   118
Sources and Nature of Allelochemicals   120
Classification of Allelochemicals   121
Allelopathy Occurrence   123
Physiological Action of Allelochemicals   124
Breeding for Ecological Allelopathic Advantage   124
Discussion Questions   125
References   126

9 EFFECTS OF STRESS ON MEMBRANES   129

Membrane Structure and Function   129
Temperature and Membrane Function   135
Ionic Interactions and Membrane Function   138
Membranes and Dehydration Stress   140
Light and Membrane Permeability   141
Membrane Permeability and Phytohormones   142
Discussion Questions   143
References   144

10 THE ROLE OF PHYTOHORMONES IN STRESSED PLANTS   145

Phytohormone Response and Water Relations   145
    Indoleacetic Acid, 146
    Gibberellin, 146
    Ethylene, 146

Cytokinins, 148
Abscisic Acid, 149
Interactions of Phytohormones in Drought Stress   151
Phytohormone Response and Temperature   152
Phytohormone Response and Nutrition   154
Phytohormone Response and Photoperiod   157
Phytohormone Response to Pathogens and Insects   162
Discussion Questions   165
References   166

**11 STRESS TOLERANCE THROUGH
    BIOTECHNOLOGY   171**

Use of Plant Growth Regulators   171
  Increasing Drought Tolerance, 172
  Cold Tolerance, 173
  Salt Tolerance, 175
  Nutrient Stress, 176
  Air Pollutants, 177
Use of Genetic Engineering   177
Stress Proteins and Tolerance   179
Discussion Questions   180
References   181

GLOSSARY   183

INDEX   195

THE PHYSIOLOGY
OF PLANTS UNDER
STRESS

1

TERMINOLOGY

It is important to understand the terminology that has evolved concerning plant stress, and we might begin with the question "What is stress?" What do we mean when we say a plant is under stress? Stress results in an aberrant change in physiological processes brought about by one or a combination of environmental and biological factors (Table 1.1). An inherent connotation in the term is that stress has the potential to produce injury. Injury occurs as a result of aberrant metabolism and may be expressed as reduction in growth, yield, or value, or death of the plant or plant part.

TABLE 1.1   Sources of Environmental Stress for Plants

Physical	Chemical	Biotic
Drought	Air Polution	Competition
Temperature	Allelochemicals (organic)	Allelopathy
Radiation	Nutrients (inorganic)	Lack of symbioses
Flooding	Pesticides	Human activities
Mechanical	Toxins	Diseases
Electrical	Salts	Insects
Magnetic	pH of soil solution	
Wind		

Equally important is the question "When is a plant not under stress?", which leads to the concept of zero stress. Zero stress is that level of exposure to an environmental factor that leads neither to injury nor to reduction in growth, yield, or value. The concept of zero stress is related to the concept of optimum conditions for growth of individual species. Variations from optimum environmental conditions might result in stress. Therefore, there are degrees of stress ranging from zero to moderate to severe and the degree of stress is related to the amount of energy entering into the change of processes within the living systems. Zero stress seldom if ever occurs but it is an important theoretical concept.

A number of stresses cause injury without producing visible symptoms. Accordingly, stress injury can occur at a subclinical as well as at a clinical level. Diagnosis of stress effects is made more difficult as a result.

Whole plants can be resistant and survive stress injury, or some parts of a single plant may be resistant (seeds, buds, dormant cells) while other parts (meristems, succulent organs, seedlings) are susceptible. Through the processes of evolution a plant species can become fit or adapted to an environment in which it thrives. Natural selection causes those paths of evolution that are successful to survive and those that are not to perish. Survivors have a tolerance of injury from environmental factors that enables them to overcome partially or completely any adverse effects. Tolerance to a stress is the capacity of a plant to survive and grow even though subjected to an unfavorable environment; the plant can sustain the effects of stress without dying or suffering irreparable injury.

Such tolerance or resistance may change as a plant grows and develops so that at one stage of development a plant may be susceptible to stress induced injury but at another stage it may be completely resistant. In addition, a stress may change metabolism, a process called acclimation, and through such changes alter the morphology and render the plant resistant to that stress. Plants that become acclimated are hardened and can survive in the new environment.

Because of the timing of development in relation to the occurrence of a stress, some plants escape injury.

Levitt (1980) divides resistance to stress into tolerance and avoidance. Stress avoidance occurs if a plant does not come to thermodynamic equilibrium with the stress or can exclude the stress by means of a physical or metabolic barrier. Stress tolerance occurs if the plant comes to thermodynamic equilibrium with the stress but no injury occurs or injury that does occur is repaired. In the process of evolution, the selection has been toward avoidance mechanisms that are more efficient than tolerance mechanisms in causing resistance to stresses. Levitt's concepts, which involve an analogy to stress–strain relationships as used in physics, have not come into widespread usage and are somewhat confusing primarily because of the difficulty in recognizing the category of strain involved.

Examples of resistance to individual stresses are:

1. *Temperature*. Plants, with few exceptions, attain the temperature of the ambient environment. They are poikilotherms. Because of this, they must have some form of tolerance to temperature stress.

2. *Drought*. Terrestrial plants are normally turgid and, thus, resistance results from avoiding loss of turgor. Drought resistance should probably be divided into dehydration avoidance or postponement and dehydration tolerance (Kramer, 1980).

3. *Irradiation*. Because of the penetrating nature of radiation, plants cannot escape it. Tolerance of irradiation stress depends upon the intensity and duration of the radiation and the amount of energy absorbed.

4. *Salts*. Plants that grow with root systems in soil of high salt content have low osmotic potentials as a result of an increased concentration of solutes and are salt tolerant. Some plants are resistant because they have mechanisms by which they exclude salt or by which the salt is concentrated in vacuoles.

5. *Nutrient deficiency*. Tolerance of deficiency of a nutrient may depend on the ability of the roots to exude metabolites enabling them to obtain more of the nutrient from the soil, as with iron, or to use substitute ions such as sodium for potassium.

6. *Nutrient toxicity*. Much of the tolerance of toxicity is the result of the processes of exclusion at the root or by concentration in vacuoles of leaf cells. Either process prevents the toxic concentration from interfering with metabolism.

The reaction of plants to stress conditions may be pathological. It has been estimated that more than 50% of all plant diseases are caused by improper environmental, nutritional, or physical conditions. Such diseases are distinguished from those caused by infectious organisms or viruses and are referred to by terms such as noninfectious, abiotic, or physiogenic. Organisms that cause diseases in plants are called pathogens. Perhaps the term physiogens should be used for those conditions of the physical environment that cause disease and the process of development of such diseases is physiogenesis as distinguished from pathogenesis. The study of the physiology of plants under stress thus becomes the study of physiogenesis.

Additional terminology is included in the glossary. Because your concepts may change, you are encouraged to further develop the description of the terms as your knowledge of stress physiology increases and your circumspection of the terminology broadens. Understanding terminology is the key to understanding a new subject and in the rapidly expanding subject of stress physiology there is still disagreement on the usage of terms. We have tried to be consistant in the use of terms in the chapters that follow.

REFERENCES

Agrios, G. N. 1978. *Plant Pathology*. Academic, New York, pp. 3–5, 673–686.

Christiansen, M. N. 1982. World environmental limitations to food and fiber culture. In M. N. Christiansen and C. F. Lewis (Eds.), *Breeding Plants for Less Favorable Environments*. Wiley-Interscience, New York, pp. 1–11.

Fitter, A. H., and R. K. M. Hay. 1981. *Environmental Physiology of Plants*. Academic, New York, pp. 3–25.

Kramer, P. J. 1980. Drought stress and the origin of adaptations. In N. C. Turner and P. J. Kramer (Eds.), *Adaptation of Plants to Water and High Temperature Stress*. Wiley-Interscience, New York, pp. 7–20.

Levitt, J. 1980. *Responses of Plants to Environmental Stresses*, Vol. 1. Academic, New York, pp. 3–18.

Levitt, J. 1982. Stress terminology. In N. C. Turner and P. J. Kramer (Eds.), *Adaptations of Plants to Water and High Temperature Stress*. Wiley-Interscience, New York, pp. 437–439.

Ritchie, J. T. 1980. Plant stress research and crop production: The challenge ahead. In N. C. Turner and P. J. Kramer (Eds.), *Adaptation of Plants to Water and High Temperature Stress*. Wiley-Interscience, New York, pp. 21–29.

Salisbury, F. B., and C. W. Ross. 1978. *Plant Physiology*, 2nd ed. Wadsworth, Belmont, CA, Chaps. 23 and 24, pp. 350–374.

2

DROUGHT STRESS

Drought is a meteorological term that means a lack of precipitation over a prolonged period of time. Sometimes physiologists refer to the resulting effect on plants as water stress but water stress may also occur over relatively short periods of time. Because ecologists have used drought stress frequently in their writings it is used in this chapter with water stress as a more specific term with the understanding that sometimes it is difficult to separate the two and they are often used interchangeably.

BASIC WATER RELATIONS CONCEPTS

Much has been written about the water relations of plants and the factors affecting them. General references at the end of this chapter may serve to review the subject. The diurnal cycles of water potential and osmotic potential have been described extensively in these references. Our purpose is to build on the fundamental knowledge of the physiology of water relations and to describe the chemical and physical changes in processes in plants that result when water loss is greater than water absorbed over extended periods of time. A summary of some of the salient concepts of plant-water relations will help establish a basis of understanding.

Water comprises 85 to 90% of the fresh weight of most living herbaceous plants. In higher plants, water is absorbed by roots from soil and is translocated to the shoots as a result of pressure gradients developed either from root pressure or transpiration. Whenever the rates of water loss by transpiration exceed the rates of water absorption by roots, water in the conducting tissues is subject to a tension (negative pressure), that is, its potential is lowered and competition for water among the various tissues and organs of the plant takes place because the equilibria among the separate water potentials have been disturbed.

The water potential (Ψw) in plants is the sum of turgor potential (Ψp), osmotic potential ($\Psi\pi$), and matric potential (Ψm). The relationship is given by the equation: $\Psi w = \Psi\pi + \Psi p + \Psi m$. Osmotic potential is created by dissolved particles, either molecules or ions, and is lowered in proportion to the number of particles in solution. Matric potential is the result of water adhering to surfaces and interfaces where the molecules of water become more ordered in arrangement and give up some of their kinetic energy. Activity of the water molecules is lost and the kinetic energy may be dissipated as thermal energy.

Turgor potential is created by water molecules bombarding the surfaces of membranes and cell walls retaining water in a closed system such as a vacuole. Turgor pressures are usually positive and are opposed by the membranes, the cell walls, or hydrostatic pressure caused by gravity in columns of water in the xylem tissues. Maintenance of turgor is necessary for growth and, if turgor decreases, wilting of parts of the plant may be visible. Turgor potential is the first component of water potential to be affected by water or drought stress.

A brief review of the status of water in the soil and how water moves through the soil may be useful in understanding how water reaches the surface of the root where it may be absorbed. Soil is composed mainly of particles and aggregates of various sizes interspersed with spaces filled with gases and water vapor or, under conditions of saturation, with liquid water. As water is removed from the soil by drainage, evaporation, or absorption by roots the continuity of the liquid water is interrupted. Some water remains on the surface of soil particles and some changes into vapor in the soil pore spaces. As more and more water is removed from the soil, that which remains is more tightly held on the soil particle surfaces. One can say that the water potential becomes lower and lower until it reaches a point at which a plant root can no longer remove enough water to overcome the deficit imposed by the loss in transpiration or other water-using processes. The plant may wilt and not recover even if transpiration is stopped by placing the plant in an atmosphere of high humidity. The plant has permanently wilted and the amount of water in the soil is at the permanent wilting point (PWP). The water potential in the soil at this point is somewhere between -1.0 and -2.0 MPa (megapascals, 1 MPa = 10 bars). Frequently, crop plants exist in a soil moisture range between -0.03 MPa and the permanent wilting point (-1.5 MPa) (Figure 2.1).

Water contacts roots in two ways: (1) the water may move to the root or (2) the root may grow and intercept moist soil. Movement of water in soil is most easily accomplished at saturation and follows Darcy's law for saturated flow. Darcy's law may be stated as

$$V = K \, \frac{\text{change in total } \Psi w \text{ in soil in cm } H_2O}{\text{change in depth or distance}}$$

where K is the coefficient of hydraulic conductivity. Remember that Darcy's law is an expression for liquid water flow. As the water potential decreases to -1.5 MPa, the value of K decreases more rapidly to 10^{-3} that of the saturated soil value.

At low soil moisture content, the continuity of water films is broken and liquid flow no longer occurs. The contribution of the matric potential to water potential becomes higher in relation to the contribution of osmotic potential. Under these conditions, water vapor movement becomes important. Water may move along the surface of a soil particle, vaporize into a pore space, condense onto the surface of another soil particle, and move long distances by repeating the process. Temperature gradients in soil aid such movement and account for water vapor movement upward in winter and downward in summer.

The root–soil interface is a dynamic region. That portion of the soil that lies near the surface of the root and is influenced by chemicals and pressure from the growing root is often referred to as the rhizosphere. The rhizosphere is a dynamic ecosystem in which populations of microorganisms are often much greater than in the bulk soil. The root processes that result in exudation of chemicals, ions, and CO_2 into the soil, as well as those that result in water and ion absorption, occur here. However, the interface near the root tip is different from the interface some distance from the tip

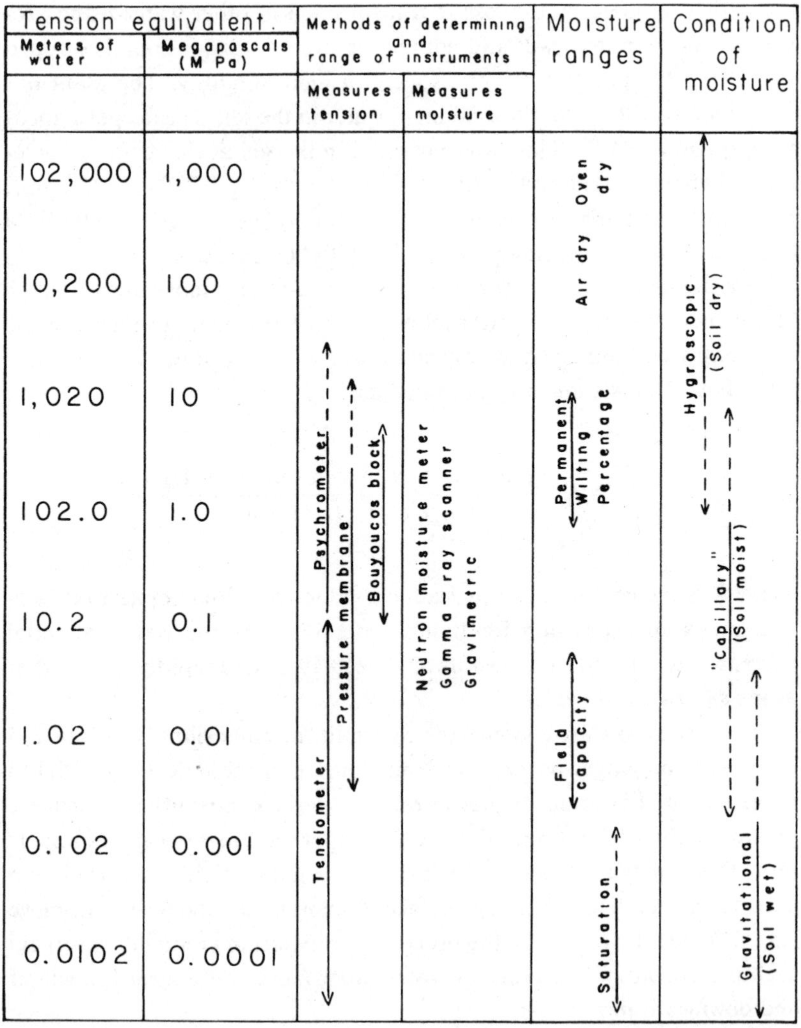

Tension equivalent.		Methods of determining and range of instruments		Moisture ranges	Condition of moisture
Meters of water	Megapascals (M Pa)	Measures tension	Measures moisture		
102,000	1,000				
10,200	100				
1,020	10				
102.0	1.0				
10.2	0.1				
1.02	0.01				
0.102	0.001				
0.0102	0.0001				

Methods of determining and range of instruments (measures tension): Psychrometer, Pressure membrane, Tensiometer. (measures moisture): Bouyoucos block, Neutron moisture meter, Gamma–ray scanner, Gravimetric.

Moisture ranges: Oven dry, Air dry, Permanent Wilting Percentage, Field capacity, Saturation.

Condition of moisture: Hygroscopic (Soil dry), "Capillary" (Soil moist), Gravitational (Soil wet).

Figure 2.1. Some soil moisture values and relationships with indicated methods or instruments of measurement.

because of differences in root growth pattern and in differentiation of surface tissues from the root tip toward the older portions of the root.

As water is absorbed from the soil near the root surface, the water potential of the soil decreases creating a gradient toward the root. At or near the root tip the zone of decreasing water potentials in the soil may be spherical in shape. At older root surfaces, where absorption is through cracks and breaks in the cortical tissues, the zones of absorption that occur around these openings are discontinuous along the root length.

Root hairs may play a role in water transport by traversing air gaps between the root and moist soil particles. Water moves many times more rapidly along a root hair than it does as water vapor diffusion across such gaps. The role of root hairs in increasing root absorbing surface for water is questionable because some evidence indicates that the resistance at the base of the root hair for water movement into the root is great enough to overcome any advantage in surface area increase.

Some roots, principally those of grasses, have a mucilage associated with the root cap. Mucilage composed of complex polysaccharides becomes inhabited by bacteria that utilize the mucilage and sloughed root cap cells as a substrate. The resulting complex, the mucigel, binds soil particles and the root tip into close association and plays a role in maintaining root–soil contact as drought stress develops and the soil volume shrinks.

The factors of root origin that affect water absorption and movement into the xylem are complex. Water enters the root, moving down water potential gradients along the root wherever there is an opening or pathway, but the most rapid absorption occurs in a region behind the root tip where the processes of elongation of cells and root hair formation occur. The pathway of least resistance occurs in the apoplast of the cortex up to the endodermis. The suberized casparian strip in the walls of endodermal cells is relatively impermeable to water. Differentiation of the endodermis may be one cause of root resistance to water movement from the root surface to the xylem. Resistances across root tissues are much higher than resistances to water movement across soil until the PWP is approached. There is some controversy about how water traverses the endodermis but most current theories involve movement into the symplast before the endodermis is reached or at the endodermis. However, movement to the xylem is primarily through the apoplast of the pericycle and into the xylem. One might view the root as an absorbing organ in which water exists in continuity from the surface into the xylem and factors that affect water potential in the xylem also affect the water potential in the rhizosphere.

It is apparent that water absorption is affected by the number of roots inhabiting a given volume of soil, the size of the roots, the rate of root length growth, the pattern of differentiation of vascular tissue within the root, the distance water must travel in the root from its site of absorption to the site of its loss or utilization, the age of the root at the absorbing site, whether the roots are perennial or annual, and the rate of new root production. These factors are discussed in Chapter 3 in relationship to drought resistance.

The water vapor loss from plant leaves has a great effect on the water status of plants and on metabolic processes. Conditions affecting evaporation rate from leaves are those that favor a steep gradient of water potential from the evaporating surface within the leaf to the atmosphere. Thus, high temperature of the ambient atmosphere coupled with a low relative humidity favor high rates of transpiration. Within limits, air movement also increases the gradient. Together, these conditions contribute to the transpiration potential. Whether the potential is reached or not depends upon the available water and such plant factors as resistance to flow through the plant and the stomatal aperture. When the transpiration potential is high and the stomates are open, high rates of transpiration can be expected unless available water is limited. In the latter situation, the plant is placed under stress.

Drought stress occurs when available water in the soil is reduced and atmospheric conditions cause continued loss of water by transpiration and evaporation. Stress may occur on a daily basis or over a prolonged period. If continued stress occurs, plants may die of desiccation unless they possess mechanisms of resistance whereby water loss is either prevented or slowed in certain tissues and organs, or unless they are able to increase rates of absorption and translocation of water.

As a drought situation continues and there is no precipitation to replenish the water in the soil, plant stress increases. Progressive development of stress was analyzed by Slatyer (1967) for a transpiring plant in a nearly saturated soil that was allowed to dry over a period of several days (Figure 2.2). In addition to a daily cycle of changes in leaf and root water potentials, there is a decline in soil water potential. The plant cycle results from a lag in water absorption compared to water loss through transpiration. At the permanent wilting point, the water potential of the leaves remains below the water potential of the soil. At this point, water can no longer move to the roots rapidly enough to overcome the deficit in the plant even though the stomates are closed. One can say that the permanent wilting

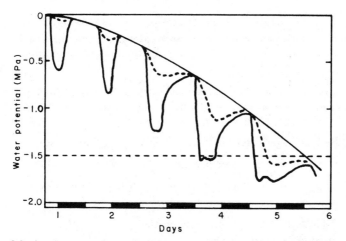

Figure 2.2. A schematic representation of the changes in leaf, root surface, and bulk soil water potentials associated with the decrease of available water in the soil. Upper solid line is the water potential of the soil. Lower solid line is the water potential of the leaf. Broken line is the water potential of the root. Permanent wilting occurs during Day 5. From R. O. Slatyer, 1967. *Plant Water Relationships.* Used with permission from Academic Press, New York.

point is controlled by the osmotic potential of the leaves and death by desiccation results as this severe stress continues.

A daily cycle of water stress results from the daily cycle of transpiration rate controlled by stomatal aperture. When the stomates are closed, usually at night except for the crassulacean plants, water abosrption continues from the available water supply in the soil. Early in the drought, the deficit of water loss over water absorption is overcome early in the night but, as drought continues and water in the soil is less available, the deficit is made up more slowly until, under severe drought, the plant does not recover, permanent wilting occurs, and eventually desiccation and death.

EFFECTS OF DROUGHT ON GROWTH AND YIELD

Drought has profound effects on growth, yield, and plant quality. The first effect of the stress may well be a loss of turgor that affects the rate of cell expansion and ultimate cell size. Loss of turgor is probably the process most sensitive to water (drought) stress. The result is a decrease of growth rate, of stem elongation, of leaf expansion, and of stomatal aperture. The terms drought stress, water stress, and water deficit are used to designate conditions within a plant when water loss exceeds water absorption and are often used interchangeably.

According to Hsiao (1973) the mechansims underlying the responses of plants to water stress may be divided into five categories:

1. Reduction of water potential or activity of cellular water.
2. Decrease of cell turgor pressure.
3. Concentration of small molecules and macromolecules as cell volume decreases with reduced turgor.
4. Alteration of spatial relations in the plasmalemma, tonoplast, and organelle membranes by volume changes.
5. Change in structure or configuration of macromolecules by removal of water of hydration or through modification of structure of adjacent water.

Hsiao (1973) chose to designate mild stress as lowering Ψw of leaves only tenths of a megapascal and lowering the relative water content (RWC) as little as 8 or 10%. Moderate stress occurs if Ψw is lowered to -1.2 to -1.5 MPa and RWC more than 10% but less than 20%. Plants under severe stress have a Ψw lowered more than -1.5 MPa and RWC more than 20%. If the RWC is lowered more than 50%, that is, if plants lose more than half of their tissue water, one can say they are desiccated. Of course, such designations are purely arbitrary, but they do help in discussing the status of the stress at a point in time.

Moderate water stress affects translocation indirectly by altering the source to sink relationships for assimilates. For example, because cell expansion is reduced, the source (leaves) is smaller and less photosynthate is available for translocation to fruits. On the other hand, the sink (fruits) size may be reduced, and if drought stress occurs after leaf expansion, the result is that the competition between leaves and fruits is lessened. Some of the relationships between the timing of drought stress and plant development are outlined in Figure 2.3. Effects on yield depend upon which part of the plant is harvested.

As stress develops, a competition for water occurs within the plant.

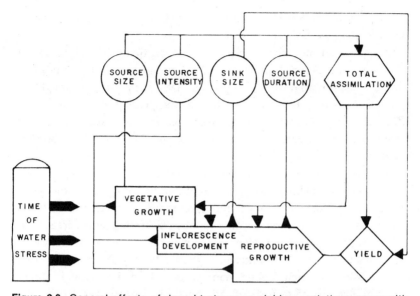

Figure 2.3. General effects of drought stress on yield as variations occur with time in the physiological processes that precede yield alterations. Processes affected are carbon dioxide assimilation, translocation (source to sink relations), and the ontogenetic development of the plant. Source size is usually the leaf size that is effected by drought, source intensity is the rate of photosynthesis in leaves, source duration is the length of time that photosynthesis continues before leaf fall or before there is a significant reduction in rate as drought continues. Dependent upon which part of the plant is harvested, the effect of the duration and time of the drought relative to the development of the plant could affect yield differently. From T. C. Hsiao, E. Ferreres, and D. W. Henderson, 1976. Water stress and dynamics of growth and yield of crop plants. In O. L. Lange, L. Kappen, and E. D. Schultze (Eds.), *Water and Plant Life: Problems and Modern Approaches.* Used with permission of Springer-Verlag, Berlin.

What factors control the distribution of water? One factor is the stage of growth and development of various plant parts. Meristems may compete favorably because synthesis of cellular material, that is, protein, creates a matric potential that contributes to lowering the water potential and the establishment of a strong Ψw gradient toward the meristematic tissue. Vacuolation also requires water. Those parts of the plant in which solute content is increasing, for example, photosynthesizing leaves, are able to compete favorably for water when plants are under mild stress. Organs in which insoluble organic material is converted to soluble organic material may also compete favorably for water.

In plants that have sizable canopies, the location of the plant part in relation to its exposure to incident radiant energy and to the atmosphere plays a role in internal water competition. Direct exposure to sunlight, as opposed to shade, affects temperature of the plant part, which in turn affects the water gradients. Water loss from shaded leaves may be less because of the increased relative humidity in the atmosphere brought about by reduced temperature.

The timing of drought stress conditions in relation to the stage of plant development is also important in terms of internal competition for water. Water stress is especially critical during reproductive development. Fruits and grains may not enlarge because rapidly transpiring leaves create lowered water potentials in the xylem, which may result in water actually leaving the fruits. Of critical significance is the availability of water at the time of grain filling in cereals. What happens in upland rice plants is a good example. The last 55 to 60% of the growth is characterized by the four developmental phases consisting of panicle initiation, production of gametes, anthesis and fertilization, and grain filling. Water deficits occurring at any of these crucial stages causes loss of yield. Perhaps the gamete production stage and anthesis are the most sensitive. Reduction in number of flower primordia occurs if water deficits occur at panicle initiation and results in lower spikelet numbers per panicle. Stress at anthesis or gamete production means infertility can occur with an increase in percentage of unfilled grains. Stress during grain filling produces lower weight of individual grains (O'Toole and Chang, 1979). Drought stress of relatively short duration, as much as 7 days to 2 weeks, may not have a drastic effect because of the difference in time of development of tillers in the field and because of a redistribution of water stored in the plants (relative water content) at the time the drought occurs. The effect of osmotic adjustment on turgor maintenance in developing reproductive structures and meristems is discussed in Chapter 3.

As a result of reduction or loss of turgor there may be a reduction in cell size accompanied by a reduction of leaf area. Reduction in leaf area as a result of drought stress has secondary effects because of a reduction of the irradiated surface area. There is an accompanaying overall reduction of the amount of photosynthates produced, which in turn contributes to the reduction of growth. Smaller leaves, in addition to reducing carbon assimilated, also reduce water loss.

The reduction of leaf expansion occurs at threshold leaf water potentials that vary with species. For example, in sunflower, leaf water potentials of -0.2 to -0.4 MPa cause cessation of leaf expansion (Boyer, 1970), in corn the threshold leaf water potential is -0.7 MPa (Acevedo et al., 1979), and in soybean it is -1.2 MPa.

EFFECTS ON ULTRASTRUCTURE

Under mild to severe drought stress there is a disruption of compartmentation with the concomitant release of hydrolases that act on substrates that are normally protected by isolation in compartments. The acid and alkaline lipases that are released destroy membranes in plants susceptible to stress, while in plants resistant to stress the membrane systems remain intact and a recovery system allows repair of what little damage is done. There is, however, a "point of no return" in both susceptible and resistant plants that, if exceeded, results in death. After reaching the point of no return, rehydration aggravates the damage. Under mild stress in corn the tonoplast may lose structure, releasing the vacuolar contents. Chloroplasts burst when they come into contact with the vacuolar sap and do not recover activity upon rehydration. In jack bean (*Canavalia ensiforme*), water stress slowed the greening process in developing leaves but the site of the block of chlorophyll synthesis was not elaborated.

Several lines of evidence indicate that there are changes in the photosynthetic mechanism in the chloroplasts and that a change in the ultrastructure may occur under water stress. When wheat plants were subjected to water stress the reduction of photosynthetic rate was not directly proportional to the stress imposed (Wardlaw, 1967). The reduction in stomatal aperture accounted for only about one half of the reduction in photosynthesis and even when CO_2 was forced into the leaf, photosynthesis did not increase (Redshaw and Meidner, 1972). When the techniques became available to isolate individual chloroplasts from leaf tissue it was demonstrated that free chloroplasts showed reduced photosynsthesis at Ψw of -0.8 to -1.2 MPa (Boyer and Bowen, 1970).

Some chloroplasts, such as those from sunflower, are irreversibly altered at -2.5 MPa, probably because of changes in the chloroplast membranes. In contrast, the chloroplasts from sorghum were still able to function after they were stressed to -3.7 MPa and then removed from the stress.

EFFECTS OF WATER STRESS ON PHOTOSYNTHESIS

The effects of water stress on photosynthesis are not well understood. Under water stress conditions that cause closure of stomates, radiant energy continues to be intercepted and absorbed by leaves. Reducing substances are produced that ordinarily would be used in reducing the carbon of CO_2 but with stomates closed, CO_2 entrance into the leaf is limited. What then, happens to the reductant?

With limited CO_2 available, the reducing compounds must either be stored and accumulate or be used. The subsequent use of carbohydrate reserves results in a deficiency of carbon compounds for both growth and maintenance processes.

In C_3 plant metabolism the first product of CO_2 fixation is 3-phosphoglyceric acid (PGA), a three-carbon compound (Figure 2.4). In C_4 metabolism the first products of CO_2 fixation are four carbon organic acids

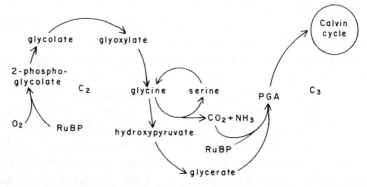

Figure 2.4. Photorespiration in C_3 plants occurs partly in the peroxisomes and partly in the mitochondria. There is a competition in the pathways that is dependent on the CO_2 and the O_2 concentrations. High oxygen concentration inhibits carbon dioxide fixation and reduces photosynthesis. If carbon dioxide concentration is lowered, as in water-stressed plants when stomates close, photorespiration increases. PGA, phosphoglyceraldehyde; RuBP, ribulose bisphosphate oxygenase.

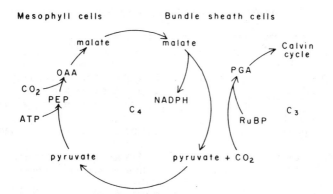

Figure 2.5. An example of the relationship between C_3 and C_4 metabolism in a malate forming C_4 plant such as maize. In some C_4 plants aspartate serves the role that malate does in this example. From M. F. Hipkins, 1984. Photosynthesis. In M. B. Wilkins (Ed.), *Advanced Plant Physiology*. Used with permission of Pitman Publ. Co., London. ATP, adenosine triphosphate; NADPH, reduced nicotine adenine dinucleotide phosphate; OAA, oxaloacetic acid; PEP, phosphoenolpyruvate; PGA, phosphoglyceraldehyde; RuBP, ribulose bisphosphate oxygenase.

(Figure 2.5). Other characteristics that differentiate between the two types of metabolism are listed in Table 2.1. There are not only metabolic differences but also structural differences in the leaves of the two types of plants. Illuminated leaves of C_3 plants evolve CO_2 in photorespiration at rates above 20% of the net photosynthesis rate and are, as a result, less efficient than C_4 plants. Dark respiration rates of mature leaves do not change markedly during mild stress so the difference in rate of photorespiration between C_3 and C_4 plants becomes an important factor in survival during drought stress.

If one uses measurements of CO_2 concentrations as an indication of photosynthetic rates, the CO_2 compensation concentration at which CO_2 leaving the leaf just equals that entering the leaf is 50 μL CO_2/L air for C_3 plants and less than 5 μL CO_2 for C_4 plants. The difference is caused by the difference in photorespiration rates in the two kinds of plants.

Under drought stress, the CO_2 exchange of the illuminated leaves is altered by changes in stomatal aperture and by differences in resistance to diffusion of CO_2 in the mesophyll cells. Since photorespiration in C_3 plants does not decrease as rapidly as gross photosynthesis, under severe stress photorespiration rates remain high. The leaf under these conditions uses

TABLE 2.1   A Comparison of Some of the Characteristics between C$_3$ and C$_4$ Plants[a]

C$_3$ Plants	C$_4$ Plants
Ps inhibited by increasing oxygen levels above 2%	Oxygen has little effect on Ps increases to high irradiances
Saturates at light intensities above 1/3 full sunlight	Pr is small, insensitive to oxygen or temperature
Compensation conc. is 50μL CO$_2$/L air	Compensation conc. is less than 5μL CO$_2$/L air
Illuminated leaves in air evolve CO$_2$ in Pr rates from 20 to 75% of net Ps	Leaf cells visibly, functionally differentiated into distinct tissues
Mesophyll cells	Different enzyme complement
Thin cytosol	Bundle sheath, small cells with
Large central vacuole	small vacuoles and many
Not visibly, functionally differentiated	chloroplasts, mitochondria, peroxisomes. Closely
Many temperate climate plants	adpressed to vascular tissue
Tolerate temp. from 5 to 30°C	Loosely packed mesophyll cells
Lower productivity	with large central vacuole,
Lower water efficiency	scattered chloroplasts, cells between bundle sheaths
	Tolerate temp. 15 to 40°C
	High rates of productivity

[a]Ps, photosynthesis; Pr, photorespiration.

carbon sources within the leaf instead of synthesizing them, with the net result of death from starvation if the severe stress endures. Although C$_4$ plants are able to survive longer because of lower photorespiration rates, they too eventually succumb to severe drought stress from causes other than starvation.

Of course, temperature interacts with the effects of water stress on photosynthesis but the interactions have been little studied. The same holds true for measurements of the oxygen concentration within the leaf. Because low oxygen reduces photorespiration rates in C$_3$ plants, it is possible to experimentally reduce the differences in the rates of photorespiration between C$_3$ and C$_4$ plants by changing the oxygen concentration supplied, but this probably does not occur under field conditions.

A third group of plants, mostly succulents and arid region plants, normally have closed stomates during the day and fix CO$_2$ in the dark at night when the stomates are open. Malic acid is synthesized and stored until

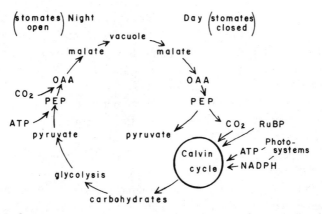

Figure 2.6. Crassulacean acid metabolism (CAM) in which C_3 metabolism is separated from CO_2 fixation by time and not by location as in C_4 metabolism. ATP, adenosine triphosphate; NADPH, reduced nicotine adenine dinucleotide phosphate; OAA, oxaloacetic acid; PEP, phosphoenolpyruvate; RuBP, ribulose bisphosphate oxygenase.

light energy is available to generate photosynthetic reducing power that fixes the CO_2 released from malate decarboxylation. Such a metabolic schematic for these crassulacean acid metabolism (CAM) plants is shown in Figure 2.6. If the stomates were open during the day, photosynthesis would proceed by the C_3 pathway. At least one plant, *Mesembryanthemum crystallinum*, shifts from C_3 to CAM metabolism as the growing season advances and the conditions change from mesic to xeric in the normal climatic cycle.

It has been estimated that CO_2 fixation is about 40% less efficient in C_3 plants than in C_4 plants but there may be some advantages of C_3 metabolism for survival (Laing et al., 1974). The recycling of CO_2 generated from photorespiration in C_3 plants (Figure 2.4) uses the reductant from the Hill reaction to refix the carbon. Such a cycle may have limited utility in stressed C_3 plants. Alternative pathways for the energy dissipation are fluorescence or reactions with oxygen. In either event, large reserves of carbohydrate are consumed.

In C_3 plants the photosynthetic carbon reduction cycle is linked to the photorespiring carbon oxidation cycle by the activity of ribulose diphosphate carboxylase–oxygenase. The proportion of carbon flowing is determined by the relative concentrations of CO_2 and O_2. In C_4 plants the mesophyll cells lack the ribulose diphosphate carboxylase–oxygenase enzyme and many of the enzymes of the photosynthetic carbon reduction and photorespiratory carbon oxidation pathways. All the enzymes are contained in the bundle sheath cells, however, where low O_2 levels may prevent photorespiration. What processes act as sinks for the photochemical energy from the Hill reaction when the stomates are closed under water stressed conditions? It is not clear at this time whether or not metabolites from the bundle sheath cells function as a sink.

There appears to be a CO_2 homeostasis, which is an important feature of plants that adapt to a slowly developing water stress. Processes come into play that help maintain a constant intercellular concentration of CO_2 when leaves are illuminated under water stressed conditions. The slowly adapting features of both stomatal and nonstomatal components of the photosynthetic system are involved. Under rapidly developing stress, which seldom occurs in the field, the system does not adapt and photoinhibition occurs in the chloroplasts (Osmond et al., 1980).

NITROGEN METABOLISM UNDER WATER STRESS

The distribution of carbon to metabolites in drought stressed leaves has been studied using $^{14}CO_2$. Although less of the carbon enters stressed leaves because of reduced stomatal aperture, it has been possible to ascertain that more $^{14}CO_2$ goes to certain amino acids than to organic acids and sugars. In unstressed corn, for example, more of the fixed carbon is in organic acids than in amino acids, but with mild stress more of the carbon appears in amino acids. Glycine and serine increase. Under severe stress the competition for carbon skeletons between alanine synthesis and the glycolate pathway increases with more of the carbon going to glycine synthesis, which causes an increase in concentration.

Nitrate reductase activity declines in water stressed leaves. The decrease may be related to a lowered translocation of nitrate in the xylem (Shaner and Boyer, 1976a, b). Still unclear is whether the change in nitrate reductase activity is the result of loss of enzyme activity, decreased rate of enzyme synthesis or an increased rate of enzyme degradation.

Free proline may accumulate in plants under water stress. Boggess et al. (1976) proved that the increase in proline is the result of synthesis de novo and not protein degradation. It is doubtful that proline accumulation contributes to the survival of plants even though it does contribute some solute to the osmoticum (Hanson et al., 1977).

Betaine accumulation has a similar pattern to proline in relation to stress. According to Hanson and Nelson (1978) the betaine arises as a result of de novo synthesis from one- and two-carbon precursors during water stress. Under drought stress barley leaves can accumulate betaine at the rate of 200 nM per 10-cm leaf per day upon wilting. The betaine is synthesized from serine.

Protein synthesis is also affected by water stress. Even under mild stress there is a shift from polyribosome content to monoribosomes that are inactive in protein synthesis (Hsiao, 1973).

WATER LOGGING AND ANAEROBIOSIS

The primary effect of flooding of soil is the creation of an anaerobic environment from which plants, and especially plant roots, cannot obtain oxygen. The diffusion rate of oxygen in water is approximately 10^{-4} that of air and consequently the rate of replenishment to plant surfaces that are submerged is extremely low. Soils at field capacity have 10 to 30% of the volume composed of air-filled spaces but the percentage decreases as water content increases. Under most conditions the oxygen supply in the air-filled spaces is replenished from the atmosphere at rates more than sufficient to maintain concentrations of oxygen in the range of 15 to 20%; under flooded conditions, however, the oxygen supply is greatly reduced.

In marshes and swamps and along the shores of oceans, lakes, rivers, and ponds, the pore spaces become saturated with water. Lack of oxygen causes death of terrestrial plants. Although some anaerobic respiration may occur in roots, the decrease in energy released from respiratory substrates is insufficient to maintain vital processes. The rates of water and nutrient absorption are curtailed and death of the roots occurs if flooded conditions continue.

Studies of tomato plants (Bradford and Hsiao, 1982) showed that after 24 h of flooding stomatal conductance and translocation were reduced by 30 to 40%. Although leaf water potential remained unchanged, root conductance decreased. The production of ethylene caused petiole epinasty, which reduced the amount of light intercepted by the leaves. Three hypotheses were proposed to explain the effect of flooding on stomatal con-

ductance: the water logged roots could export some factor that caused closure of stomates; the roots could fail to export sufficient quantities of some
factor normally present in xylem sap that prevents stomatal opening; or
there could be a reduction in phloem transport to anaerobic roots, which
could lead to a buildup of assimilates or growth regulators in the leaves.

Flooding also caused closing of stomates in seedlings of the woody species, *Fraxinus pennsylvanica* (Koslowski and Pallardy, 1979). There was
some evidence of stomatal reopening after a critical period of flooding.
Upon removal of the flooding stress, stomates opened within 6 to 10 days.
The reaction of *Ulmus americana* seedlings (Newsome et al., 1982) was
similar but there was no stomatal closure in leaves that expanded during
the flooding. The formation of adventitious roots after flooding occurred
alleviated the effects of injury to the roots present before the flooding. Tolerance to flooding occurs during dormancy of roots for some woody species
(Coutts and Philipson, 1978; Coutts, 1981).

Symptoms associated with flooding injury are:

1. Pattern of yellowing leaves from base to top of plant
2. Drooping of the petioles while plant is still turgid
3. Leaf epinasty
4. Hypertrophy (a swelling of cells)
5. New root formation from the stems
6. Wilting under severe conditions of flooding

Ethylene production has been implicated in the production of some of
the symptoms of flooding injury. For example, Drew et al. (1979) linked
ethylene production to adventitious root and aerenchyma formation in
corn plants subjected to reduced aeration. Both of these reactions increase
tolerance of flooding conditions. Kawase (1974) was able to cause flooding
symptoms on sunflower by applications of ethephon, a chemical that
releases ethylene when metabolized by the plant. (See also Chapter 10.)

DISCUSSION QUESTIONS

1. What is drought stress? What are the primary symptoms of drought
    stressed plants?

2. Assuming that water stress is brought about by a transpiration rate greater than the water absorption rate, what are the factors that affect the severity of stress? How do they operate?

3. What is the role of root hairs in the absorption of water under conditions of drought stress?

4. Would resistance to water movement through the root affect the development of stress symptoms? Qualify your answer.

5. One of the earliest events that occurs in plants susceptible to water stress is reduced turgor. What are some possible ways turgor reduction can bring about changes in metabolism characteristic of water stress?

6. The relationship between drought stress and growth is a complex one. What are some of the physiological concepts one must consider in an explanation of reduced growth and yield caused by lack of water?

7. Water stress produces altered metabolism. What changes occur in the processes of photosynthesis and nitrogen metabolism as the degree of stress progresses from mild to severe? Are there accompanying ultrastructural changes?

8. The following algebraic expression represents a quantitative expression of one of the factors in the soil-plant-air continuum that can cause plant water stress. How does it relate to the development of plant water stress? What other factors should be considered?

$$\text{Water potential of the air} = \frac{RT \ln \text{rh}/100}{V}$$

where RT is the gas constant $\times$ absolute temperature, rh is the relative humidity, and V is the molal volume of water $= 18$.

REFERENCES

Acevedo, E., E. Fereres, T. C. Hsiao, and D. W. Henderson. 1979. Diurnal growth trends, water potential, and osmotic adjustment of maize and sorghum leaves in the field. *Plant Physiol.* 64:476-480.

Barnett, N. M., and A. W. Naylor. 1966. Amino acid and protein metabolism in Bermuda grass during water stress. *Plant Physiol.* 41:1222-1230.

Boggess, S. F., C. R. Stewart, D. Aspinall, and L. G. Paleg. 1976. Effect of water stress on proline synthesis from radioactive precursors. *Plant Physiol.* 58:398–401.

Boyer, J. S. 1970. Leaf enlargement and metabolic rates in corn, soybeans, and sunflower at various leaf water potentials. *Plant Physiol.* 46:233–235.

Boyer, J. S., and B. L. Bowen. 1970. Inhibition of oxygen evolution in chloroplasts isolated from leaves with low water potentials. *Plant Physiol.* 45:612–615.

Bradford, K. J., and T. C. Hsiao. 1982. Stomatal behavior and water relations of water logged tomato (*Lycopersicon esculentum*) plants. *Plant Physiol.* 70:1508–1513.

Bradford, K. J., T. C. Hsiao, and S. F. Yong. 1982. Inhibition of ethylene synthesis in tomato (*Lycopersicon esculentum*) plants subjected to anaerobic root stress. *Plant Physiol.* 70:1503–1507.

Coutts, M. P. 1981. Effects of water logging on water relations of actively growing and dormant Sitka spruce seedlings. *Ann. Bot.* 47:747–753.

Coutts, M. P., and J. J. Philipson. 1978. Tolerance of tree roots to water logging. *New Phytol.* 80:63–69.

Drew, M. C., M. B. Jackson, and S. Giffard. 1979. Ethylene-promoted adventitious rooting and development of cortical air spaces (aerenchyma) in roots may be adaptive responses to flooding in *Zea mays* L. *Planta* 147:83–88.

Drew, M. C., A. Chammel, J. Garrec, and A. Fourcy. 1980. Cortical air spaces (aerenchyma) in roots of corn subjected to oxygen stress. *Plant Physiol.* 65:506–511.

Drew, M. C., M. B. Jackson, S. C. Giffard, and R. Campbell. 1981. Inhibition by silver ions of gas space (aerenchyma) formation in adventitious roots of *Zea mays* L. subjected to exogenous ethylene or to oxygen deficiency. *Planta* 153:217–224.

Hanson, A. D., and W. D. Hitz. 1982. Metabolic responses of mesophytes to plant water deficits. *Annu. Rev. Plant Physiol.* 33:163–203.

Hanson, A. D., and C. E. Nelson. 1978. Betaine accumulation and ^{14}C-formate metabolism in water stressed barley leaves. *Plant Physiol.* 62:305–312.

Hanson, A. D., and N. A. Scott. 1980. Betaine synthesis from radioactive precursors in attached, water stressed barley leaves. *Plant Physiol.* 66:342–348.

Hanson, A. D., C. E. Nelson, and E. H. Everson. 1977. Evaluation of free proline accumulation as an index of drought resistance using two contrasting barley cultivars. *Crop Sci.* 17:720–726.

Hipkins, M. F. 1984. Photosynthesis. In M. B. Wilkins (Ed.), *Advanced Plant Physiology*. Pitman, London, pp. 219–248.

Hsiao, T. C. 1973. Plant responses to water stress. *Annu. Rev. Plant Physiol.* 24:519-570.

Hsaio, T. C., E. Ferreres, and D. W. Henderson. 1976. Water stress and dynamics of growth and yield of crop plants. In O. Lange, L. Kappen, and E. D. Schultze (Eds.), *Water and Plant Life: Problems and Modern Approaches.* Springer-Verlag, Berlin, pp. 81-305.

Kawase, M. 1974. Role of ethylene in induction of flooding damage in sunflower. *Physiol. Plant.* 31:29-38.

Kluge, M. 1976. Crassulacean acid metabolism (CAM): CO_2 and water economy. In O. L. Lange, L. Kappen, and E. D. Schulze (Eds.), *Water and Plant Life— Modern Approaches.* Springer-Verlag, Berlin, pp. 313-322.

Koslowski, T. T., and S. G. Pallardy. 1979. Stomatal responses of *Fraxinus pennsylvanica* seedlings during and after flooding. *Physiol. Plant.* 46:155-158.

Kramer, P. J. 1969. *Plant and Soil Water Relationships: A Modern Snythesis.* McGraw-Hill, New York.

Kramer, P. J. 1983. *Water Relations of Plants.* Academic, New York.

Laing, W. A., W. L. Ogren, and R. H. Hagman. 1974. Regulation of soybean net photosynthetic CO_2 fixation by the interaction of CO_2, O_2 and ribulose 1,5-diphosphate carboxylase. *Plant Physiol.* 54:678-685.

Levitt, J. 1980. *Responses of Plants to Environmental Stresses, Vol. 2, Water, Radiation, Salt, and Other Stresses.* Academic, New York, pp. 25-228.

Newsome, R. D., T. T. Kozlowski, and Z. C. Tang. 1982. Responses of *Ulmus americana* seedlings to flooding of soil. *Can. J. Bot.* 60:1688-1695.

Osmond, C. D., K. Winter, and S. B. Powles. 1980. Adaptive significance of CO_2 cycling during photosynthesis in water stressed plants. In N. C. Turner and P. J. Kramer (Eds.), *Adaptation of Plants to Water and High Temperature Stress.* Wiley-Interscience, New York, pp. 139-154.

O'Toole, J. C., and T. T. Chang. 1979. Drought resistance in cereals—rice: A case study. In H. Mussell and R. C. Staples (Eds.), *Stress Physiology of Crop Plants.* Wiley, New York, pp. 374-405.

Redshaw, A. J., and H. Meidner. 1972. Effects of water stress on the resistance to uptake of carbon dioxide in tobacco. *J. Exp. Bot.* 23:229-240.

Routley, D. G. 1966. Amino acid and protein metabolism in Bermuda grass during water stress. *Crop Sci.* 6:358-361.

Salisbury, F. B., and C. Ross. 1978. *Plant physiology.* Wadsworth, Belmont, CA.

Shaner, D. L., and J. S. Boyer. 1976a. Nitrate reductase activity in maize (*Zea mays* L) leaves. I. Regulation by nitrate flux. *Plant Physiol.* 58:499-504.

Shaner, D. L., and J. S. Boyer. 1976b. Nitrate reductase activity in maize (*Zea mays* L) leaves. II. Regulation by nitrate flux at low leaf water potential. *Plant Physiol.* 58:505-509.

Singh, R. N., L. G. Paleg, and D. Aspinall. 1973. Stress metabolism I. Nitrogen metabolism and growth in the barley plant during water stress. *Aust. J. Biol. Sci.* 26:45-46.

Slatyer, R. O. 1967. *Plant-Water Relationships*. Academic, New York.

Tolbert, N. E. 1973. In D. D. Davis (Ed.), Rate control of biological proceses. *Symp. Soc. Exp. Biol.* 27:299-318.

Waldren, R. P., I. D. Teare, and S. W. Ehler. 1974. Changes in free proline concentration in sorghum and soybean plants under field conditions. *Crop Sci.* 14:447-450.

Wardlaw, I. F. 1967. The effect of water stress on translocation in relation to photosynthesis and growth. I. Effect during grain development in wheat. *Aust. J. Biol. Sci.* 20:25-39.

3

DROUGHT RESISTANCE

Drought resistance should probably be divided into dehydration avoidance or postponement and dehydration tolerance (Table 3.1). In dealing with drought resistance one must keep in mind that stress may be different in degree. As drought continues, stress to which plants are exposed may progress from mild to severe. In addition, the stage of development of the plant when the stress occurs affects the resistance to injury and affects survival.

OSMOTIC ADJUSTMENT

One of the physiological changes plants may undergo as stress develops is osmotic adjustment. Increases in the osmotic potential of cells, triggered by the stress, help the plant maintain turgor. In fact, small changes in turgor may be regarded as the most probable means by which stress is transduced to metabolism. Factors that help maintain turgor are (1) a lowering of the osmotic potential, (2) a capacity to actively accumulate solutes, (3) highly elastic cells or tissues, and (4) small cells.

As stress develops there is a gradual lowering of the osmotic potential as the solute concentration increases because of the loss of solvent water. But

TABLE 3.1   Terms Used in Drought Stress Resistance

Drought resistance	Dehydration avoidance	Ludlow (1980)
	Dehydration tolerance	
Drought tolerance	Dehydration postponement	Kramer (1980)
	Dehydration tolerance	
Drought resistance	Drought avoidance	Levitt (1980)
	Drought tolerance	
Drought escape		

beyond the normal occurrence, in some plants there is an increase of net solute content of cells. This effect can be shown by comparing what happens to the Ψp in leaves of a resistant and a susceptible cultivar of a plant such as wheat when the Ψw decreases (Figure 3.1). There are a number of other examples of this kind for irrigated compared to nonirrigated plants. Graphic plots of the change in turgor pressure against the change in water potential show that there is an accumulation of solutes in the cells of plants as they undergo stress. Only rarely are changes in solute concentration correlated with Ψw in studies that measure solute concentration. Osmotic adjustment as a physiological phenomenon has been overlooked until recently.

Factors that affect osmotic adjustment are:

1. *The rate of development of stress.* A Ψw decrease of -0.1 to -0.5 MPa per day allows for adjustment. A decrease of Ψw of -1.0 to -1.2 MPa per day is too rapid.

2. *Degree of stress.* Early in the stress cycle full turgor may be maintained but continuation of stress results in less capacity to adjust fully.

3. *The environmental conditions.* Those factors that affect the rate of drying such as temperature and light intensity have direct effects. The day and night temperatures and other factors that affect the rate of photosynthesis and production of solutes have an indirect effect.

4. *Difference in cultivars and in tissue or organ exposed.*

5. *The age of the plants.*

Osmotic adjustment can give rise to full or partial maintenance of turgor. For convenience it is usual to measure the increase in solutes either at

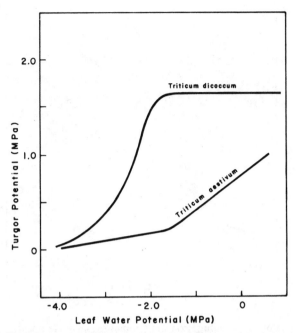

Figure 3.1 The relationship between turgor potential and leaf water potential for two wheat genotypes *Triticum aestivum* and *Triticum dicoccum*. From N. C. Turner and M. M. Jones, 1980. Turgor maintenance by osmotic adjustment: A review and evaluation. In N. C. Turner and P. J. Kramer (Eds.), *Adaptation of Plants to Water and High Temperature Stress*. Copyright © John Wiley and Sons, Inc., 1980.

a water potential of zero, at which point the turgor pressure equals the osmotic potential, or at zero turgor, at which point the turgor pressure equals the osmotic pressure. Under either of these conditions

$$\text{Osmotic potential} = \frac{\Psi \pi_0 V_0}{V}$$

where V is the osmotic volume and $\Psi \pi_0$ and V_0 are at some reference value, for example, full turgor or zero turgor. Confusion may result if the two aspects of increasing solute concentration in the cells are not kept differentiated. Other terms have been used to describe the phenomenon, such as, osmoregulation, turgor regulation, water activity regulation, and osmotic adaptation.

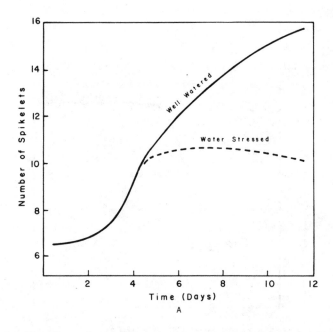

Figure 3.2 The effect of drought on the number of spikelets per apex (A), apex length (B), and lamina length of leaf number 5 (C). Water was withheld from the wheat plants on Day 0. From E. W. R. Barlow, R. E. Munns, N. S. Scott, and A. H. Reisner, 1977. Water potential, growth and polyribosome content of the stressed wheat apex. *J. Exp. Bot.* 28:909–914. Used with permission from Oxford Univ. Press, Oxford, UK.

Early in the development of drought stress full turgor may be maintained by the process of osmotic adjustment but as the stress continues there is less capacity to adjust fully.

Let us apply these concepts to what happens in the wheat apex as an example of the affect of drought stress on apical meristems and tolerance mechanisms. In the apex some tissue is enclosed within mature leaf sheaths and is protected to a degree from the atmosphere. The apex is not connected to the body of the stem by functional xylem. Both of these features reduce transpiration and the water content changes little under stress conditions, which allows osmotic adjustment to take place. It has been determined that when the apex water potential decreases by 0.5 MPa the processes of spikelet initiation (Figure 3.2A), leaf elongation (Figure 3.2B), and apex elongation (Figure 3.2C) are inhibited.

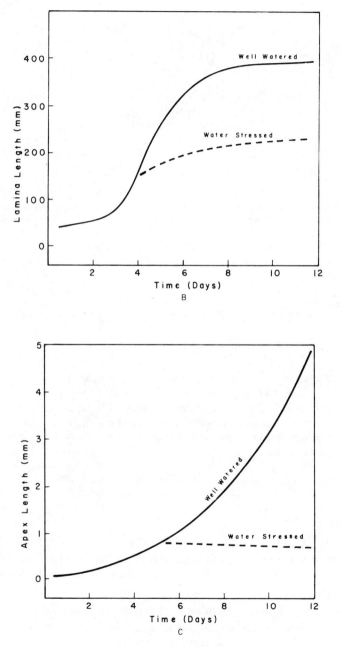

Figure 3.2 *(Continued)*

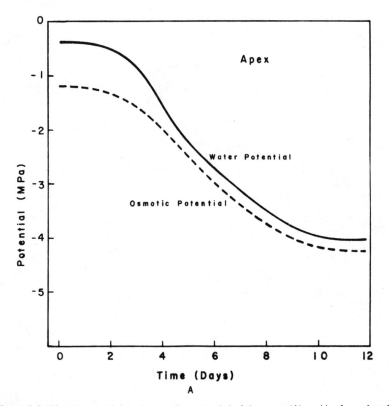

Figure 3.3 Water potential and osmotic potential of the apex (A) and leaf number 4 (B) in water-stressed wheat plants. Water was withheld from the plants on Day 0. From R. E. Munns, C. J. Brady, and E. W. R. Barlow, 1979. Solute accumulation in the apex and leaves of wheat during water stress. *Austr. J. Plant Physiol.* 6:379–389. Used with permission from the *Australian Journal of Plant Physiology.*, East Melbourne, Australia.

Survival of the wheat apex under drought stress depends on its ability to maintain turgor through osmotic adjustment as illustrated in Figure 3.3A and B in which the water potential, turgor pressure, and osmotic potential are compared with what happened in a mature, exposed leaf.

The accumulation of free amino acids and soluble sugars was correlated with the increase in resistance to the stress (Figure 3.4A and B).

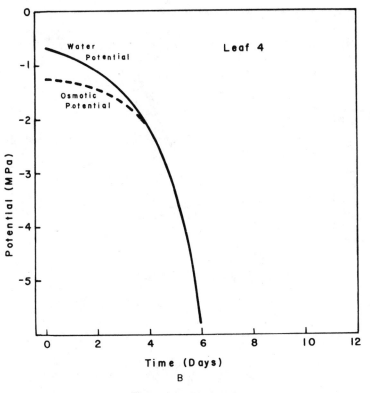

Figure 3.3 *(Continued)*

CELL ELASTICITY

There is some ambiguity about the role of elasticity in drought stress resistance. Because of the stress there is a decrease in cell size that results in an increase in elasticity. Conversely, a fourfold increase in cell size results in only a halving of the elasticity. Accompanying the increase in elasticity as the cells decrease in size is an increase in the dry weight to turgid weight ratio. In Chapter 4 there will be more discussion about elasticity and resistance to stress and the effects of freezing on cellular resistance using isolated protoplasts for experimental material.

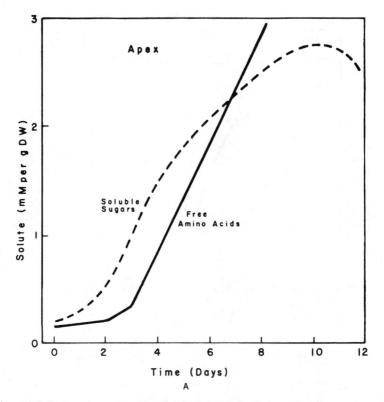

Figure 3.4 Free amino acid and soluble sugar contents in millimoles per gram of ethanol insoluble dry matter of the apex of leaf 7 (A) and of leaf 4 (B) in water-stressed wheat plants. From R. E. Munns, C. J. Brady, and E. W. R. Barlow, 1979. Solute accumulation in the apex and leaves of wheat during water stress. *Austr. J. Plant Physiol.* 6:379–389. Used with permission from the *Australian Journal of Plant Physiology*, East Melbourne, Australia.

DROUGHT ESCAPE

Plants can escape drought by various mechanisms that enable them to remain dormant during the drought or to complete their life cycle during favorable moisture conditions. Desert ephemerals, for example, grow and reproduce during periods of adequate rainfall and produce seeds that remain dormant during periods of drought. Plants with a rapid rate of development and an indeterminant growth habit are best adapted for escaping drought.

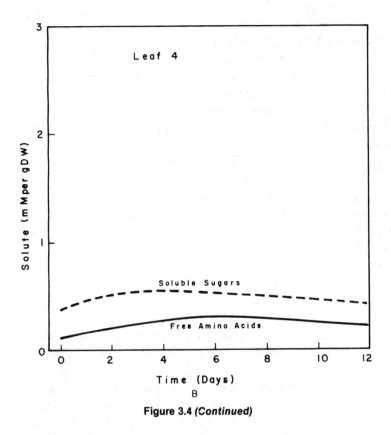

Figure 3.4 (Continued)

DROUGHT TOLERANCE

Turner (1979) has distinguished between two types of tolerance to drought:
(1) tolerance at high tissue water potential and (2) tolerance at low tissue
water potential. Each has special aspects with regard to the mechanisms of
tolerance. Tolerance at high tissue water potential results from either a
reduction of water loss or a maintenance of water uptake. Reduction of
water loss follows from an increase in stomatal and cuticular resistance, a
reduction in absorbed radiation, or a reduction in leaf area. Under
drought stress the hydrostatic feedback control of stomatal aperture over-
rides other controls such as the CO_2 concentration. For example, in the
barrel cactus the stomates closed after 40 days of stress and remained
closed for 7 months. In this time period the water potential of the cactus
fell from -0.1 to -0.6 MPa but the soil water potential fell to -9.0 MPa.

For many crop plants the leaf water potential at which the stomates close is -0.8 to -2.8 MPa. Of course, for some crop plants the stomates do not completely close. The best advantage is if all stomates close and the cuticular resistance is high. Water stress may cause development of thicker cuticle, which increases cuticular resistance.

Reduced radiation absorption arises through movement of leaves so that the angle of incidence to incoming radiation provides less area for absorption, the development of pubescence, which insulates the leaf surface, or the development of increased waxiness and reflecting qualities of the leaf surface. Reduction in leaf area comes about as the result of a reduction in number of leaves produced, or the number of tillers formed, or abscission of leaves under the influence of drought stress. The morphological adaptation of leaves to water stress may occur during growth and development of the leaves, or after they are fully developed. As a result of the reduction of turgor, leaf area may be reduced even before the reduction in stomatal aperture and photosynthesis.

A reduction in leaf area also reduces water loss. Drought stress causes an increase in rate of senescence of older leaves and premature abscission, which still further reduces transpiration. In addition, these older leaves may not be contributing much carbohydrate to developing grains or fruits. These kinds of adaptations are largely irreversible.

After full development of leaves, other adaptations may occur in addition to the change in leaf angle and surface reflectance. Leaf rolling is common in the grasses as a result of the loss of turgor in the bulliform cells along the midrib on the upper epidermis of leaves. Rolling may reduce transpiration by as much as 70% and the leaf area exposed to incident radiation by as much as 68% (Begg, 1980). The leaf adaptation mechanisms in desert plants have been subjected to an elaborate system of classification (Evanari et al., 1975).

Maintaining water uptake is the result of properties of root systems. Deeply rooted plants can continue to absorb water until the drought becomes severe and reaches deep into the soil. The rate of growth of roots may also affect the tolerance to stress. As the roots grow more, the root-to-shoot ratio changes. Increased root growth may result in less shoot growth, or the increased length and density of roots may change the ratio. Low resistance to water movement through the roots and the conducting system to the shoots by an increase in vessels or vessel diameter also aids in maintaining water uptake.

Some scientists hold the view that the true tolerance to drought stress occurs when plants are subjected to drought and the water potential is reduced. The mechanisms of tolerance in these circumstances may be the retention of high tissue turgor because of conditions that produce a low water potential or a high tissue elasticity. Low water potential is the result of a low osmotic potential probably produced through osmotic adjustment and the addition of solutes to the cell sap. The concept of tissue elasticity as it affects the maintenance of turgor is more complex and involves the relationship of turgor to cell volume. Small cells are more elastic than large cells, hence tissues made up of small cells are more tolerant of drought stress.

There is another kind of tolerance that occurs when some kinds of plants such as the lichens, mosses, and resurrection plants are almost completely dehydrated and yet recover when they are rehydrated. Such tolerance must occur at the protoplasmic level and has been termed desiccation tolerance. Some parts of crop plants may possess desiccation tolerance. For example, seeds are very resistant to drying as are some rhizomes and leafless stems. For the ressurection plants, tolerance of desiccation in the range of -160.0 MPa and below may occur. Foliage of at least 83 angiosperms can tolerate these low levels of water potential and still survive. There is some evidence that such tolerance may increase if the stress develops slowly (Gaff, 1980).

MEASUREMENT OF DROUGHT STRESS

No one method can measure all the types of resistance to drought stress. Kramer (1969) has stated rather firmly that the only reliable indicators of water stress are measurements made on the plants themselves. He has proposed four criteria to be met in developing any test: (1) There should be a good correlation between rates of physiological processes and degree of water stress. (2) A given degree of stress should produce similar physiological changes in a wide range of plants. (3) The units used and the methods of producing units measured should apply to a wide range of plants. (4) The test should require a minimum of plant tissue.

What physiological processes can be measured that reflect drought stress in the plant? The obvious ones are a decrease in turgor and the degree of stomatal aperture. Metabolic rates and disturbances of enzymatic processes that result in the accumulation of metabolites are less obvious and harder to correlate. Other criteria that are the result of longer exposure to the conditions of stress are cuticle thickness, root density, and the size of the conducting tissues from the roots to the shoots.

Although these techniques may give some useful information concerning the development of drought stress, there exists a need for the development of better criteria and techniques for these measurements. Levitt (1980) has proposed some possible reasons for the difficulties encountered in the measurement of drought stress. Many plants continue to grow during drought because of water retention caused by osmotic adjustment and increases in cuticular resistance or stomatal resistance. Dehydration tolerance occurs as a result of synthesis of isozymes and in some plants metabolic repair of injury may occur upon removal of the stress.

Yield is a poor measure of a particular stress because, in the field, yield reflects the effect of a combination of many factors. Survival time has been used in a variety of ways and has given rise to the use of terms such as the Ausdauer unit (see Levitt, 1980). An Ausdauer unit is the water content at complete stomatal closure divided by the cuticular transpiration rate. Water content is calculated by subtracting the water content at death from the water content at stomatal closure. Such a calculation gives an approximately correct relative order of resistance within a group of plants under one set of conditions but not from one population to another.

The removal of leaves or shoots and allowing them to dehydrate under standard conditions has also been used to measure resistance. The method excludes the contribution of roots to survival time and is affected by the water content at the time of excission and the rate of transpiration. Efficiency of water utilization or water requirement has been a common expression used by some agronomists for many years and has been used to separate plants into the "water savers" and the "water spenders." The calculation involves dividing the grams of dry weight produced by the kilograms of water used to produce the dry weight. A variation of the measurement is to divide the net efflux of water by the net influx of carbon dioxide over a time period.

Another concept involves the stress at which 50% of the plants are killed and measures the critical drought avoidance of the plant. The water potential of the environment at which 50% of the plants are killed is divided by the water potential of the plant at the 50% killing point. But perhaps the most meaningful method is the measurement of the energy level of water in the stressed plants. Many concepts have been evolved over the years to express the energy level. The water deficit or water saturation deficit concept involved the measurement of field or fresh weight of whole leaves and the turgid weight of leaves standing in water in a humid chamber. The turgid weight minus the fresh weight was divided by the turgid weight minus the dry weight to obtain the water deficit. A variation of this was to use leaf discs for the weight determinations. With the advent of the thermocouple psychrometer, water potential can be measured directly, the osmotic potential determined, and turgor calculated by difference. Various plasmometric techniques explained in plant physiology texts also may be used to measure the osmotic quantities and, hence, the energy of water in stressed plants. The pressure bomb of Scholander et al. (1965) has been used successfully as a quick method of measuring xylem water potential under stress conditions.

Sullivan (1971) makes three recommendations for use in measuring resistance: (1) desiccation tolerance tests or heat tolerance tests give information on how much tissue drying can occur before injury; (2) field measurements of water potential or relative water content show how far the internal water status is kept above the critical point during drought; and (3) diffusive resistance or stomatal aperture indicate if internal water potential is kept up by retarded transpiration or whether efficient root absorbing and water conducting systems are keeping the shoot supplied with water.

BREEDING FOR DROUGHT STRESS RESISTANCE

Many sources of information may give hints to breeders about the most useful characteristics for determining drought resistance. Some of these are:

1. Observations from a single genotype.
2. Comparisons among species, races, or cultivars from contrasting environments.
3. Interventions in the metabolism of one genotype by metabolic blocks or reinforcements of some metabolic reaction.

4. Comparisons between drought resistant and drought susceptible cultivars.

5. New approaches to look at proteins synthesized under stress conditions. (Stress proteins are discussed in Chapter 11.)

6. Use of mutants that lack one or another characteristic. Those traits or characteristics that appear to be most useful and practical at the present time are earliness to maturity, root growth, stomatal control, stomatal number, cuticular resistance, cell turgor, and some chemical traits (Quizenberry, 1982).

Some of the sources of information have proven ineffective and impractical when applied to crops in the field so that new approaches and tools are needed for identifying resistant progeny in segregrating populations.

If a crop has a short life cycle, it may escape the drought stress that would affect a crop with a longer life cycle, but the advantage occurs only in some environments. Upland cotton has higher yields if earlier maturing varieties are used and drought occurs. Productivity in these varieties may actually be reduced if rainfall occurs or if there is above normal moisture. Each day of earlier maturity in winter wheat has also been shown to impart a yield advantage. Depending on the crop, escape from rust diseases, insects, and summer heat stress can occur.

Increased root growth increases the volume of stored soil moisture available to the plants and increases resistance to drought. As indicated before, the density and extent of root development correlates with drought resistance. Larger root systems relative to shoot growth are favorable to resistance but it may be that reduced root development during the vegetative growth may result in more water available to fill grain later in the life cycle of plants such as wheat. Increased root growth may also use part of the total photosynthate available and may reduce the growth of the shoots. Although selections can be made for these traits with the result of higher yields in crops such as barley, wheat, and sorghum, the selection process in segregating populations has not proven practical.

Plants use less than 5% of the water that passes through them but it is important that the water be conserved. Crop plants vary from -0.8 to -2.5 MPa in the amount of stress needed for stomatal closure. What is the water content of the plant when stomates close? It depends on evironmental conditions such as the number of cycles of previous drying. For stomatal closure as a trait, parental lines can be detected only during the midday stress and selection is possible for midday stomatal closure. It is a dominant character in cotton and possibly in sorghum and soybeans as well.

Selection for stomatal closure is useless unless selection for cuticular resistance also occurs. A high cuticular resistance prevents cuticular transpiration.

The number of stomates per unit of leaf surface varies among genotypes of a species and is genetically controlled. In beans and corn, low stomatal frequency has been correlated with high rates of photosynthesis, but in barley photosynthesis was not correlated with stomatal frequency, so the effects vary.

Since the decrease in water potential of the leaf over time is not linear with the turgor pressure, cell turgor is not useful in predicting yield. Besides, osmotic adjustment may occur, which helps the plants retain their turgor.

There have been some attempts to correlate with resistance the increase in concentration of certain chemicals that accumulate during stress, Such chemicals are proline, betaine, and abscissic acid. The accumulations are not consistent with the differences in productivity so these traits are not yet useful in a breeding program.

DISCUSSION QUESTIONS

1. Distinguish between drought tolerance and drought avoidance as mechanisms of drought resistance, using appropriate examples of each.

2. You are working with a plant breeder who is trying to develop a drought-resistant crop plant for use in a semiarid region. What would be your advice:
   a. as to which traits to include in the ratings and screenings?
   b. as to how to measure and compare drought resistance for the progeny?

3. Assume that you are to establish criteria for a good test for resistance to drought stress. What would be four of the most important criteria?

4. What factors enter into drought resistance?

5. Under what conditions does one measure drought avoidance most meaningfully?

6. Compare plant reactions that lead to resistance to drought stress to those necessary for survival. Are they different? Refer to specific processes whenever possible.

7. Over the years the question of how some kinds of plant species can survive desiccation has plagued scientists. What are some of the hypotheses for protoplasmic tolerance to this severe stress that are being investigated now? Why is it important to know?

8. As stress develops, which processes or parameters are affected first? Which ones are affected late in the development of stress?

REFERENCES

Acevedo, E., E. Fereres, T. C. Hsiao, and D. W. Henderson. 1979. Diurnal growth trends, water potential, and osmotic adjustment of maize and sorghum leaves in the field. *Plant Physiol.* 64:476-480.

Barlow, E. W. R., R. E. Munns, and C. J. Brady. 1980. Drought responses of apical meristems. In N. C. Turner and P. J. Kramer (Eds.), *Adaptation of Plants to Water and High Temperature Stress*. Wiley-Interscience, New York, pp. 191-205.

Begg, J. E. 1980. Morphological adaptations of leaves to water stress. In N. C. Turner and P. J. Kramer (Eds.), *Adaptation of Plants to Water and High Temperature Stress*. Wiley-Interscience, New York, pp. 33-42.

Bewley, J. D. 1979. Physiological aspects of desiccation tolerance. *Annu. Rev. Plant Physiol.* 30:195-238.

Brown, R. W., and B. P. van Haveren. 1972. Psychrometry in water relations research. Utah Agr. Exp. Sta., Utah State University, Logan.

Cavalieri, A. J., and J. S. Boyer. 1982. Water potentials induced by growth in soybean hypocotyls. *Plant Physiol.* 69:492-496.

Christiansen, M. N., and C. F. Lewis. 1982. *Breeding Plants for Less Favorable Environments*. Wiley-Interscience, New York.

Cutler, J. M., K. W. Shahan, and P. L. Steponkus. 1980. Dynamics of osmotic adjustment in rice. *Crop Sci.* 20:310-314.

Cutler, J. M., K. W. Shahan, and P. L. Steponkus. 1980. Alteration of the internal water relations of rice in response to drought hardening. *Crop Sci.* 20:307-310.

Evanari, M., E. D. Schultze, L. Kappern, U. Buschborn, and O. L. Lange. 1975. Adaptive mechanisms in desert plants. In F. J. Vernberg (Ed.), *Physiological Adaptation to Environment*. Intext, New York, pp. 111-130.

Gaff, D. F. 1980. Protoplasmic tolerance of extreme water stress. In N. C. Turner and P. J. Kramer (Eds.), *Adaptation of Plants to Water and High Temperature Stress*. Wiley-Interscience, New York, pp. 207-230.

Hsiao, T. C. 1973. Plant responses to water stress. *Annu. Rev. Plant Physiol.* 24:519-570.

Kramer, P. J. 1969. *Plant and Soil Water Relationships: A Modern Synthesis*. McGraw-Hill, New York.

Levitt, J. 1980. Measurement of drought avoidance. In *Responses of Plants to Environmental Stresses*, Vol. 2. Academic, New York, Chapter 6.

Michelena, V. A., and J. S. Boyer. 1982. Complete turgor maintenance at low water potentials in the elongating region of maize leaves. *Plant Physiol.* 69:1135-1149.

Quizenberry, J. E. 1982. Breeding for drought resistance and plant water use efficiency. In M. N. Christiansen and C. F. Lewis (Eds.), *Breeding Plants for Less Favorable Environments*. Wiley-Interscience, New York.

Scholander, P. F., H. T. Hammel, E. D. Bradstreet, and E. A. Hemmingway. 1965. Sap pressure in vascular plants. *Science* 148:339-346.

Shone, M. G. T., and A. V. Flood. 1980. Studies on uptake and loss of water by barley roots in relation to changes in root resistance. *J. Exp. Bot.* 31:1147-1159.

Sullivan, C. Y. 1971. Techniques for meauring drought stress. In K. L. Larson and J. D. Eastin (Eds.), *Drought Injury and Resistance in Crops*. Crop Sci. Soc. Am. Spec. Publ. No. 2, Madison, WI, pp. 1-16.

Turner, N. C. 1979. Drought resistance and adaptation to water deficits in crop plants. In H. Mussell and R. C. Staples (Eds.), *Stress Physiology in Crop Plants*. Wiley-Interscience, New York.

Turner, N. C., and P. J. Kramer (Eds.). 1980. *Adaptation of Plants to Water and High Temperature Stress*. Wiley-Interscience, New York.

Turner, N. C., and M. M. Jones. 1980. Turgor maintenance by osmotic adjustment: A review and evaluation. In N. C. Turner and P. J. Kramer (Eds.), *Adaptation of Plants to Water and High Temperature Stress*. Wiley-Interscience, New York, pp. 87-103.

4

TEMPERATURE STRESS

Plants are poikilotherms; that is, they assume the temperature of their environment. The stresses imposed by temperature have important implications for agriculture. We have historically accepted rather than addressed the effect of small climatic changes on plants. For example, it has been conjectured that a 1°C decrease in the world mean temperature would result in a 40% reduction in rice production. Imagine, if you will, what a 2°C increase in frost hardiness of citrus trees, winter cereals, potatoes, deciduous fruit tree blossoms, and tender vegetables would do to increase world yields. Similarly, a 2°C increase in hardiness of wheat might extend production to areas now in spring wheat. Other examples could be cited to show that temperature is a major uncontrollable climatic factor and point up the need for research on inheritable characteristics for temperature stress resistance and on the manipulations of crop plant physiology to increase tolerance of temperature extremes.

Current research effort can be divided into the following topics, in each of which there is a need for more knowledge:

1. Mechanisms of freezing avoidance and tolerance.
2. Effective cryoprotectants.

3. Biochemical changes in nucleic acids, polyribosomes, amino acids, sugars, and ultrastructural comparisons.

4. Changes that occur in membranes that cause them to rupture, to alter structure leading to protoplast lysis, to bring about the disruption of active transport mechanisms, and to change lipid phases that affect membrane fluidity.

5. Acclimation processes involving changes in proteins and the effects of antimetabolites, surfactants, regulators, and other solutes on acclimation processes.

6. Supercooling events occurring primarily in woody species. What kinds of ice nucleators occur and what are the genetic differences in degree of super cooling?

7. Survival traits and breeding for the control of hardiness.

Temperature stress on plants can be divided into the effects of temperatures that cause chilling injury, freezing injury, and high temperature injury. Also involved are the mechanisms bringing about hardiness or tolerance of temperature stresses.

CHILLING INJURY

Some plants, such as those of tropical origin, are injured when the temperature drops to some point above freezing but low enough to cause damage to tissues, cells, or organs of the plant. For many sensitive plants this happens when they are exposed to temperatures of about 10°C. Direct injury may take the form of any of a number of events such as necrosis, discoloration, tissue breakdown and browning, reduced growth, or failure to germinate in the case of seeds. More indirect and subtle events such as reduced grain set, delayed harvest, reduced photosynthesis, and reduced water absorption may occur as well.

On the ultrastructural level a description of what happens in the tomato hypocotyl cells upon exposure to low temperatures may prove helpful in understanding the altered physiology. Upon exposure to 5°C for 3 days, severe damage occurs at all levels of cellular organization. The membrane systems are disorganized and 1 cell in 10 dies. The effects are progressive with time. After 2 h there are only small changes. After 4 h there is loss of definition of thylakoids and the mitochondrial membranes show discontinuities. After 8 h there is less structure in the cytoplasm and the ribosomes are altered. After 12 h the changes in the membranes are drastic and the cell walls become opaque with some swelling. After 16 h all symptoms become more drastic and are accompanied by the precipitation of protein in the vacuoles. These events were described by Ilker et al. (1979).

EFFECTS ON MEMBRANES

In both chilling and freezing injury the membranes of cells and tissues are of pivotal importance. Because it is difficult to apply conventional biochemical and biophysical methods to frozen cells determination of the temperature at which injury and death occurs cannot be precise. In both chilling and freezing there appears to be a critical temperature below which injury occurs. The freeze-induced dehydration cannot be the cause of injury during chilling. A number of methods have been employed to show clearly that there is a membrane transition that occurs as the temperature changes and it is the transition that results in injury. The membrane hypothesis to explain chilling injury implies that cellular membranes in sensitive plants undergo a physical–phase transition from a normal flexible liquid–crystalline to a solid–gel structure at the temperature critical for chilling injury. Apparently there is some molecular reordering in the membranes and a change in fluidity occurs (see Chapter 10). The role of membrane proteins in the sensing of low temperature and injury is not understood well enough to satisfy all the evidence.

THE FREEZING PROCESS

A brief review of the physical chemistry of water and the freezing process is needed to understand what happens when plants are subjected to freezing temperatures. At 0°C there is a phase transition from solid to liquid or from liquid to solid. Dilute solutions, such as are found in plant cells, undergo melting or freezing at different temperatures. As the phase separation occurs, both liquid and solid coexist. The melting points of the solutions are lowered to below 0°C. The solutions usually supercool before actual freezing occurs and the magnitude of the supercooling varies with the influence of a number of factors.

Freezing may be triggered by ice nucleators, which are materials that cause crystallization of the water. Ice crystals themselves can be nucleators and the process of crystallization, once started, may proceed rapidly until a supercooled solution is completely frozen. Usually ice nucleation of water occurs between 0 and $-10°C$ but, if water occurs in small droplets of less than 10 μm diameter, the water may supercool to $-38.1°C$, which is the homogeneous nucleation temperature, ΔT_h. If the melting point depression temperature is ΔT_m, then $\Delta T_h = (-2 \Delta T_m + 38.1)$. For most plants ΔT_m is -41 to $-47°C$. Whole plants usually supercool only a few degrees but parts of some plants may supercool to the limit. Since the solutions in the cells are more concentrated than the water or the dilute solutions outside the cells or plasmalemma, a phase separation occurs during exposure to freezing temperatures.

What are the events that occur as plants or parts of plants freeze? If one starts with the temperature decreasing below the freezing point of water then:

1. Water in the cells and in the intercellular spaces supercools.

2. Ice forms extracellularly because the solute concentration is lower and there are more effective ice nucleators present.

3. Intracellular nucleation is prevented by the plasmalemma.

4. Since there is no contact of ice with intracellular water, a vapor pressure gradient is formed from inside to outside.

5. An equilibrium is established either by the water evaporating out of the cell to the extracellular ice or by the formation of intracellular ice.

6. The way in which the equilibrium is established depends on the rate of cooling in relation to the permeability of the plasmalemma and the surface area to cell volume ratio.

7. The amount of dehydration of the protoplasm that occurs depends on the intracellular osmolality. Whether the maximum amount of water is removed or not depends upon the permeability of the plasmalemma and the surface area available for efflux. If the flux is not

adequate, the equilibrium is established by intracellular ice formation, which is usually lethal to the cell.

8. Some liquid water or solution is not frozen.

9. Injury becomes apparent only upon thawing and some of the injury may be the result of the thawing process.

At equilibrium the difference in Gibbs free energy across the phase separation boundaries must be zero. There are a number of factors that affect the balance of the Gibbs free energy and contribute to the complexity of the reaction. There may be a rise in temperature of the supercooled tissue when it freezes after nucleation, shifts in concentration of the solute in the liquid at the ice interface, shifts in the density (concentration) of water in the gas phase at the interface, and a number of other factors of a physical–chemical nature.

FREEZING INJURY TO MEMBRANES

In freezing injury to membranes research involves the change in the cellular membranes when cell volume contracts because of dehydration caused by the freezing process. If a membrane can be envisioned as a mono- or bilayer of lipid, when the surface is reduced there is a tangential pressure induced within the interface. The potential energy stored in this fashion eventually exceeds the hydrophobic forces holding the layer together and lipid is lost to the environment. This suggests that the point at which tangential pressure causes irreversible loss of membrane material coincides with the limit of reversible plasmolysis. In tolerant plants the lipid is stored in the cytoplasm and then restored to the membrane upon rehydration of the cells when they are thawed. Williams et al. (1981) present evidence supporting the hypothesis that in winter wheat (a) osmotic stress to the membrane is the primary cause of cell freezing injury acting through a tangential stress in the membrane; (b) cell injury results from an irreversible loss of membrane material during plasmolysis as a result of membrane stress and cell rupture occurs upon deplasmolysis; (c) at least in Kharkov winter wheat, tolerance of freezing is caused by initial loss of a specific component of the membrane that can be stored in intracellular sites and recaptured by the membrane upon rehydration and deplasmolysis. Failure to recapture the component may help explain why protoplasts that have been plasmolyzed during freezing do not fully recover their original volume (Steponkus and Wiest, 1979).

EFFECTS OF FREEZE-THAW CYCLES ON PLASMA MEMBRANES

Freeze-thaw cycles can be used on isolated protoplasts of rye (Steponkus and Weist, 1979) to study the effect on plasma membranes of contraction and expansion (Figure 4.1). Since the contraction and expansion are caused by dehydration and rehydration, the effects of freeze-thaw cycles can be predicted by osmotic manipulations. Volumetric behavior is largely determined by osmotically active solute content and the concept of a minimum volume is equated with an osmotically inactive volume. From studies of this type it has been learned that a lytic lesion occurs that is a function of surface area of the plasma membrane during expansion. The freeze-thaw injury is a result of two strains: (1) a contraction induced membrane alteration that decreases the maximum critical surface area, and (2) an expansion-induced dissolution of the plasma membrane that occurs when the maximum critical surface area is exceeded. There is still little knowledge of what happens to the molecular structure of the membranes during the contraction and expansion processes that could be used to explain why the protoplast does not resume the same volume after expansion that it had before contraction. What is the nature of the lesion? What happens when tissues in the intact plant freeze? In addition, plasmolysis has been used as an indicator of the stresses that occur when cells are frozen. The use of plasmolysis has made it possible to examine the cellular events underlying injury by freezing dehydration and to determine the nature of resistance to injury by hardened plants (Siminovitch, 1981). The situation becomes more complicated with the evidence from freeze fracture experiments in which the membrane protein is implicated as aggregation of protein particles occurs in membranes of the frost resistant *Solanum acaule* but not in the membranes of the frost susceptible *Solanum tuberosum*. The aggregation occurred up to 10 days after the beginning of the acclimatization process of *S. acaule* but then became redistributed almost to the control level after 15 days. No such change happened in *S. tuberosum* (Toivio-Kinnucan et al., 1981).

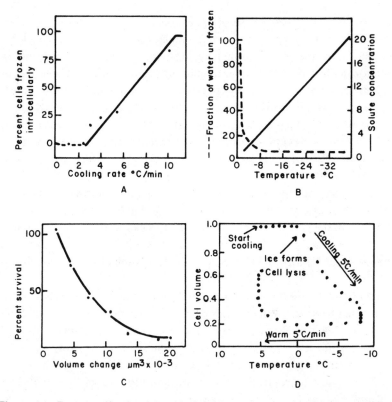

Figure 4.1. Factors affecting injury during freeze–thaw cycles as demonstrated with isolated rye protoplasts. (A) The probability that cells will freeze intracellularly at various cooling rates. (B) A schematic representation of changes in cell volume and solute concentration at decreasing freezing temperatures. (C) Cell survival as a function of cell volume change when cells previously caused to contract with various osmolalities of salt solutions were caused to rehydrate by diluting the solutions. Survival is also a function of surface area. (D) Changes in cell volume during a freeze–thaw cycle when cooling and warming rates were 5°C/min. Adapted from P. L. Steponkus, and S. C. Wiest, 1979. Freeze–thaw induced lesions in the plasma membrane. In J. M. Lyons, D. Graham, and J. K. Raison (Eds.), *Low Temperature Stress in Crop Plants: The Role of the Membrane*. Used with permission from Academic Press, New York.

The presence or absence of sterols in the membrane also has an effect on the changes that take place during the freezing process (see Chapter 10).

Tolerance to chilling has been linked to the degree of unsaturation of membrane lipids, other lipid factors, and soluble proteins or enzymes. Because there is no resynthesis of lipids in older tissues, any change in the degree of unsaturation occurs through molecular reorientation. As plants harden under the influence of lowering temperatures there may be an increase in the degree of unsaturation of the younger tissues. The degree of unsaturation has been linked to the temperature at which a phase transition occurs in the membranes of the cells. As the degree of unsaturation increases to 60% the phase transition point is lowered. Beyond 60% unsaturation there is a rapid lowering of the phase transition point. The presence of sterols such as cholesterol may shift the phase transition temperature downward.

Since sulfhydryl groups on the membrane proteins have been implicated in tolerance to chilling, it has been suggested that increased reducing power from the formation of lipid peroxidases causes a lowering of reductant and an increase in reducing power to compensate. There is also a change in the lability of enzymes with the formation of more stable isozymes.

Palta and Li (1978) have formulated an hypothesis for a sequence of events that leads to death of cells during freezing and thawing cycles (Figure 4.2). They linked the breakdown of membranes to the inactivation of potassium and sugar pump mechanisms, which leads to a change in cellular metabolism and death of the cell. Death may also be brought on by the lowered oxygen concentration as a result of infiltration of tissue by water when membrane integrity is destroyed.

TOLERANCE OF FREEZING STRESS

To understand the nature of the tolerance of freezing stress it is necessary to have some knowledge of the sources of injury that may occur. Historically, there has been much emphasis placed on the formation of intracellular ice crystals that disrupt the structure of the protoplasm and on the effect of the type and number of ice crystals that form. Of more significance may be the injury caused by dehydration of the protoplasm as the water evaporates to the extracellular ice. As water leaves the cell the con-

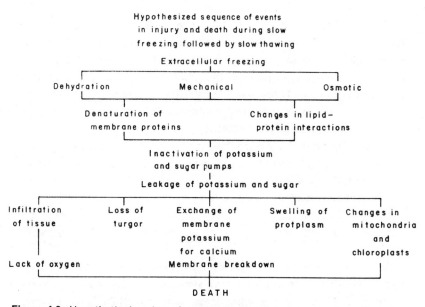

Figure 4.2. Hypothetical series of events that lead to injury and death during freeze-thaw cycles. Adapted from J. P. Palta, and P. H. Li, 1978. Cell membrane properties in relation to freezing injury. In P. H. Li and A. Sakai (Eds.), *Plant Cold Hardiness and Freezing Stress: Mechanisms of Crop Implication*. Used with permission from Academic Press, New York.

centration of solutes increases to the point where some of them may become toxic. For example, some semipolar compounds, such as phenylalanine, valine, leucine, and others, disturb the lipid–lipid interactions leading to the destabilization of membranes.

Injury upon rehydration of cells may result from the adhesion of portions of the protoplasm because of cross bonding with the result that rupture may occur. The rate of cooling and of thawing also affects the severity of injury. Factors that are involved in determining the type of injury that occurs include the duration of the freeze, the degree of supercooling, the number of cycles of freezing and thawing, and a number of conditions of the plant such as the stage of growth and development, anatomy of the tissue, moisture content of the tissue, and the growth habit of the plant. All of these factors alter either the freezing process or the sensitivity of the plant to the stress.

Plants with little or no freezing tolerance, such as corn, cucurbits, beans and others, are subject to frost damage. The foliage becomes flaccid and water soaked. Examination of the tissue usually reveals that compartmentation of the cells is destroyed and the membranes are leaky. External temperatures need be no lower than -1 to $-3°C$. There may be slight supercooling, but when nucleation happens intracellular freezing of susceptible tissues of these types of plants occurs.

Plants with limited tolerance (hardiness) withstand some ice formation. This group of plants includes spring wheat, peas, potatoes, and cabbage. Plants with the growing points below the surface of the soil or snow cover are somewhat protected. Plants in this group seldom supercool more than -1 to $-2°C$ unless the water content is low. Intracellular ice formation is rare. Death results from dehydration as water moves to the intercellular ice.

Woody plants are capable of deep supercooling. Most deciduous forest species and fruit tree cultivars have some tissues that may supercool to about $-40°C$. Let us use apple trees as an example. During summer growth, the tissues are sensitive and behave like susceptible herbaceous species. During the winter, changes have occurred and the bark and bud tissues may freeze extracellularly. Water from the cambium and phloem migrates to the cortex where it is intercellularly frozen. The xylem ray cells undergo deep supercooling. The buds in the dormant state also deep supercool.

Some woody plants do not supercool and must have other mechanisms for tolerating subzero temperatures. Tolerance in these species, many of which occcur in the arctic and subarctic climates, must be on a protoplasmic level. There may be similarities between low temperature tolerance and the protoplasmic desiccation tolerance discussed in Chapter 3.

FROST RESISTANCE AND COLD HARDINESS

Frost resistance may not be the correct term to use because of the poikilothermic state of plants. Rather, frost tolerance should be the term used. The tolerance is inducible and may be brought about by a number of possible mechanisms. All of the following have been implicated in tolerance.

1. There may be a decreased concentration of potentially toxic compounds so that, when the solutes become concentrated, they are nontoxic.

2. The ratio of nontoxic compounds to toxic compounds may be such that those that might be toxic are rendered noneffective by dilution.

3. Special "protective" compounds may shield membranes from toxic compounds.

4. There may be a decrease in the sensitivity of the membranes to toxic compounds.

5. The colligative protection of solutes such as sugars and amino acids may prevent injury.

6. There may be synthesis of soluble proteins that in some way protect the cells from injury.

EFFECT OF TEMPERATURE ON ROOT PROCESSES

The effect of varying root temperature on growth of shoots follows the general pattern shown in Figure 4.3. A number of changes occur in the physiology of the roots that may account for effects on shoot growth. A decrease in root temperature decreases water absorption rates for a number of reasons. As the temperature of water decreases its viscosity increases because there is an increase in hydrogen bonding. A decrease in temperature also decreases permeability of membranes to water. Since gases are more soluble in water at lower temperatures, there is an increase in dissolved CO_2 and O_2 that decreases the pH, which in turn reduces water absorption. Activity of solute decreases and thus there is a higher osmotic potential in the cells. All of these events may cause a water absorption lag and a water stress on plants. Root temperature at which transpiration is greatest varies among species.

Nutrient absorption is also affected by temperature (Figure 4.4). At lower temperatures there is a reduction in the amounts of nutrients released into the soil solution from parent material so the supply of nutrients is less. The active transport systems in root cells are reduced by lower temperature as are the translocation and assimilation processes. Another

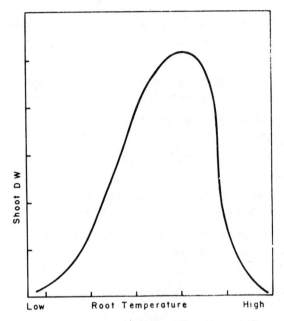

Figure 4.3. Generalized illustration of the effect of root temperature on the growth of shoots. From A. J. Cooper, 1973. *Root Temperature and Plant Growth, A Review.* Used with permission from the Commonwealth Agricultural Bureaux, Farnham Royal, UK.

factor that affects the absorption of nutrients is the reduction in the activity of microorganisms and the release of nutrients.

Translocation in plants subjected to low temperatures exhibits changes in pattern and amount of nutrients moved. The reduced root growth means that there is a reduced sink for photosynthates, which may then remain in the stems or leaves, causing the leaves to become thicker and increase in dry weight. Lower temperatures cause a reduction in synthesis and translocation of growth substances produced by the root. These substances take the form of cytokinins, amino acids, and certain vitamins.

High temperature of the root increases nitrate nitrogen content because of a reduced activity of nitrate reductase as well as an increased uptake of nitrate. Uptake by barley, for example, was increased 10 fold when root temperature was raised from 13 to 25°C.

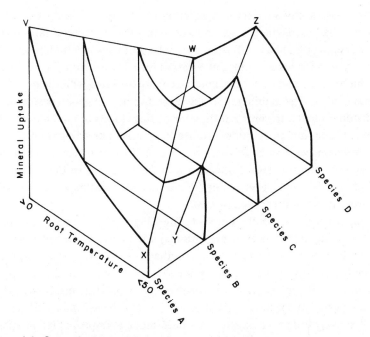

Figure 4.4. General relationship between root temperature and mineral content of plant tissue. Curves A, B, C, and D indicate four representative species. In species A the mineral content decreases with increasing root temperature over most of the range between 0 and 40°C. In species D there is an optimum root temperature at which mineral concentration is greatest. The response curve moves between these two extremes. The high concentration at low temperature decreases along line VW, and a high-temperature peak develops along line YZ. Low concentrations fall along the line XW. Mineral content is directly proportional to the rate of mineral uptake and inversely proportional to the rate of growth. From A. J. Cooper, 1973. *Root Temperature and Plant Growth, A Review.* Used with permission from the Commonwealth Agricultural Bureaux, Farnham Royal, UK.

BREEDING FOR TEMPERATURE TOLERANCE

Use of mitochondrial activity of several alfalfa genotypes within a variety has shown that there is a broad distribution of a mitochondrial phosphorylative efficiency (ADP-to-oxygen ratio) and respiration rates per unit of tissue. When plants were grouped according to these parameters and used as parents, segregation revealed that those progeny that coupled efficient oxidative phosphorylation with low respiration rates had higher yields and better tolerance of summer temperatures (McDaniel, 1982). Those progeny that had exceedingly high or exceedlingly low respiration rates died.

Marshall (1982) has outlined some of the problems involved in developing a system for breeding for temperature tolerance in crop plants. One is the inefficiency and high experimental error associated with the use of the freezing tests. The inability to screen large numbers of genotypes, to select individual plants, to control testing for distinct forms of possible freezing stress, has limited the usefulness of freezing tests in breeding programs. If the problems were to be solved, precise and efficient freezing tests would circumvent the problems associated with field testing. In the absence of effective measurement techniques it is somewhat meaningless to consider improvements in breeding systems.

Experiences with bulk composite breeding in cereal crops have been discouraging in that no cultivars with major increases in the upper level of cold tolerance have resulted to date (Gullord et al., 1975). There appears to be insufficient genetic diversity for the development of new polygenic systems, or, if the necessary genetic diversity is present, insufficient time and opportunity for development of new systems have been available.

Relatively little specific effort has been made to improve high temperature tolerance through breeding and selection. The breeding task is difficult because of limited understanding of the genetic and physiological bases of heat tolerance, the compounding of heat and drought tolerance stresses, and again the lack of efficient and accurate laboratory tests to select for the physiological traits that result in heat tolerance. Knowledge of the genetic diversity for heat tolerance is limited. Sullivan (1972) has used electrolyte leakage from leaf discs as a measure of heat injury. The results correlate well with other more difficult measurements such as those of soluble protein content, enzyme resistance, and heat stability of photosynthesis in the field.

EFFECTS OF HIGH TEMPERATURE

There are situations in which plants may reach a high temperature with resulting injury. For higher plants, the emerging seedlings are subjected to the high temperature of the soil, which has absorbed infrared radiation from the sun and exceeds the tolerance of the emerging plant part. The injury that occurs is frequently mistaken for damping off. Fleshy fruits or vegetative parts of plants may have a buildup of temperature inside because the dissipation of energy is not efficiently carried out. When leaves are in high light intensity and are under water stress, the evaporative cooling that would normally reduce the temperature does not occur and leaf temperatures may be several degrees above the ambient temperature. Those parts of a plant that are exposed to direct solar radiation have higher temperatures than the shaded side of the same part of the plant. The most obvious cause of temperatures in the injurious or lethal range is a high atmospheric temperature such as occurs in deserts or in hot springs.

Symptoms of high-temperature injury are the appearance of necrotic lesions, particularly on stems and hypocotyls, a chlorotic mottling of leaves, fruits, and needles (of conifers), and death. The causes of these symptoms are varied. Because of the elevated temperature, respiration rate increases and photosynthetic rate decreases; the tissue may deteriorate because of the lack of life supporting energy. There are also alterations in the proteins or enzymes and membranes. (Heat proteins are discussed in Chapter 11.) Enzymes have specific ranges of temperature at which they are active. The denaturation and aggregation of proteins may also occur. With the disruption of membranes there is a destruction of the compartmentation of cells, enzymes come into contact with substrates that they ordinarily would not, and reactions occur that alter ion, water, and organic solute movement. The differential temperature lability of proteins (enzymes) causes changes in the basic metabolism of cells at high temperatures.

Lipid behavior, particularly that in membranes, changes with a rise in temperature. At low temperatures there is solidification of fatty acids with increasing chain length or decreasing number of saturated bonds. A rise in temperature therefore causes a change in the viscosity of the lipid components of the cells to a more fluid state. There is some evidence that high-temperature tolerance can be increased by the application of chemicals that increase the ratio of saturated to unsaturated fatty acids.

High-temperature tolerance is usually expressed as the temperature required to produce a defined level of injury or a defined level of dead plants (usually 50%). Other criteria used are plasmolysis of the protoplasm, cytoplasmic streaming, and photosynthetic rate. Tolerance of high temperatures stems from the ability to maintain high rates of photosynthesis, the stability of proteins that must resist conformational changes at high temperatures, the ability to repair or synthesize structural protein, and the possession of substances that protect the proteins from change.

There is evidence that the lower the hydration of protoplasm the greater the hydrogen bonding and the greater the tolerance of high temperatures. Thus, dormant tissues, such as in seeds and buds, are more tolerant than rapidly growing tissues. A reduction in nitrogen fertilization, therefore, produces more tolerance. Cross linkages by divalent cations may also cause more tolerance.

Some plants avoid high-temperature injury because of the angle and arrangement of the leaves toward the incident radiation. Coloration and reflectance properties of the leaves are important. Desert species may reflect as much as 70% of the incident infrared radiation. The presence of a thick cuticle and leaf hairs alters the reflectance properties of the surface and improves conductive cooling. Factors that improve the evaporative cooling of transpiration when water is available also improve the high-temperature tolerance.

TEMPERATURE ACCLIMATION

Temperature acclimation can mean development of tolerance to injury from either high or low temperature. Hence, studies of types of injury caused by temperature stress aid in determining which processes or structures might be involved in acclimation processes. We have seen previously how plants may be injured and why they are resistant to temperature stresses. Now we are ready to study processes and development of structures that change with time of exposure to stress and result in tolerance to the temperature stress. The term acclimation has been used in the past to describe the result or development toward tolerance. Practical plant scientists use the term "hardy" to describe tolerant plants and the process is called temperature hardening.

Evidence for the effects of temperature on acclimation comes from several types of experiments: (1) plants are exposed to a gradual or slow change in temperature and the changes in processes are followed throughout the temperature change; (2) clones of some species are grown at two different temperatures and reactions leading to acclimation of processes, such as photosynthesis, are recorded; or (3) species from different ecological temperature regimes are brought together under the same temperature regime and comparisons in reactions are observed. Of course the time of exposure to the temperature regimes must be controlled because the reactions of plants differ depending on the length of exposure to any one temperature. Reactions also are related to the developmental stage of the plants or plant parts.

Controlled environment facilities having both low and high temperatures are frequently used. Plants taken from regions of relatively high temperatures can be compared with plants of the same or different species taken from regions with relatively low temperatures. Also plants may be exposed to elevating temperatures over various periods of time so that they can adjust within the the limits of their genetic composition.

ACCLIMATION OF PHOTOSYNTHESIS

One would suspect that high temperatures would reduce photosynthetic rates and at the same time increase respiration rates so that the amount of photosynthate for assimilatory processes would be reduced. A more detailed explanation is possible with recent advances in biochemical methods of analysis and studies of enzyme systems.

Comparisons of photosynthetic characteristics of cold-adapted and heat-adapted species have been made. Plants from the cold-adapted species have a much higher photosynthetic capacity at low temperatures than plants from the heat-adapted species of deserts and, conversely, the desert species have a much higher photosynthetic capacity at high temperature than the cold-adapted species (Figure 4.5) (Berry and Bjorkman, 1980). The decrease in photosynthesis is attributed to an irreversible inactivation of photosynthesis at different temperatures for the two types of plants. The effect is on the photosynthetic machinery directly and not on other secondary processes that result in changes in membrane permeability and breakdown. Acclimation to temperature can involve two distinct changes: (1) the low-temperature acclimation from an increased capacity of a limiting temperature-dependent catalytic step and (2) the high-temperature acclimation from an increased stability of one or several chloroplast components (Berry and Bjorkman, 1980). Investigation of many of the enzymes involved in carbon dioxide fixation and photosystem II activity indicates that ribulose bisphosphate carboxylase and fructose diphosphate phosphatase are probably the limiting steps in the carbon metabolism cycle at

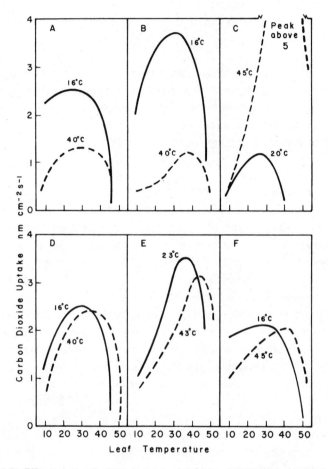

Figure 4.5. Effect of growth temperature on the rate of light-saturated net carbon dioxide uptake for several C_3 and C_4 species native to habitats with contrasting thermal regimes. Some show acclimation to either a high- or a low-temperature regime. (A) *Atriplex glabriuscula*; (B) *Atriplex sabulosa*; (C) *Tidestroma oblongifolia*; (D) *Atriplex hymenelytra*; (E) *Atriplex lentiformes*; (F) *Larrea divaricata*. Redrawn from O. Bjorkman, M. R. Badger, and P. A. Armond. 1980. Response and adaptation of photosynthesis to high temperatures. In N. C. Turner and P. J. Kramer (Eds.), *Adaptation of Plants to Water and High Temperature Stress*. Copyright John Wiley and Sons, Inc., 1980.

low temperatures and the heat stability of the thylakoid membranes is a key component in the acclimation to high temperatures.

Raison et al. (1980) have expressed a hypothesis for the reacton of membranes to temperature changes. The response to a temperature shift may be a vertical movement of proteins in the bilayer toward the aqueous interface at low temperature and toward the interior of the bilayer at high temperature. Under both conditions there would be considerable alteration in the forces maintaining the tertiary structure of the proteins.

HIGH-TEMPERATURE ACCLIMATION

When one thinks of high-temperature stress one usually thinks of thermophilic organisms that live in hot springs or of plants living in deserts. Crop plants, however, frequently encounter high temperatures during the growing season even in temperate climates. Those who do any gardening know that seasonal crops such as broccoli, lettuce, and peas grow better in the spring and fall than in midsummer. Yields may be reduced but the plants do not necessarily perish.

Photosynthesis is one of the most heat sensitive processes affecting growth and plants differ greatly in their potential for photosynthetic acclimation to temperature (Bjorkman et al., 1980). Usually, plants reflect the temperature regimes of their native habitats in the acclimation processes. The evolutionary process of natural selection may account for at least part of such reactions. But what are the mechanisms that render some plants tolerant of high temperatures and others nontolerant? Some investigators believe that it is the thermal stability of chloroplasts based on properties of the thylakoid membranes and the soluble enzymes outside of the membranes. Others believe that there may be several mechanisms of adaptation to high temperature that involve induction of protein synthesis or altered protein function. Those plants that do not adapt may be incapable of carrying out a structural or functional change in protein (Teeri, 1980). The changes in protein that occur and the adaptations to high temperature remain poorly understood.

In addition to changes in protein, there may be changes in the lipids of membranes. Raison et al. (1980) state that the loss of physiological function of membranes at both high and low temperatures may be explained on the basis of a model of a balance between relative strengths of hydrophobic and hydrophilic interactions among proteins, lipids, and the aqueous environment of the membranes.

ACCLIMATION TO LOW TEMPERATURE

Enzymatic reactions obey the laws of kinetics and as the temperature becomes lower, the rates of reactions decrease. How much the rates decrease and at what temperature they cease entirely depends upon the enzymes involved and their location in metabolic pathways. The flow of energy is of key importance in survival of cells and tissues and those reactons that are involved in energy transfer may be the key reactions involved in hardening of plants to low temperature stress.

Photochemical reactions are not temperature dependent so the production of ATP and NADP may continue at temperatures approaching freezing. When the enzymatic processes of carbon dioxide fixation are reduced or cease because of the lower temperature, there may be a shift in the photosynthetic system from carbohydrate production to cyclic phosphorylation and protein synthesis. If growth ceases, as it does in seeds and perennial plants that become dormant, the materials that usually go to assimilation of new cellular material accumulate and add to the shift toward protein synthesis that becomes a part of the low-temperature hardening processes.

Under lowering temperatures photosynthetic rates decrease but the rates of respiration and assimilation decrease even more. The photochemical reactions may not decline unless there is an accumulation of end products with an accompanying feedback loop causing a reduction in the reactions. The net result of these events is the accumulation of reducing power in the form of reduced NADP. Interruption of the use of reduced NADP causes only cyclic phosphorylation and this leads to protein and phospholipid synthesis which requires only ATP. If growth ceases under reduced temperatures, the products that would be used in growth also accumulate. One of the changes often noted in development of hardening is the increase in soluble protein.

As Levitt (1980) has pointed out, the difficulty in researching the hardening process is that each factor studied can be demonstrated to be a limiting factor and correlated with resistance to low temperature. Such events as gelation of structures, increase in volume of protoplasts, rearrangement of organelle structure, expansion of the endoplasmic reticulum, and increase in soluble protein have been associated with low temperature hardening (Trunova, 1982). The available literature suggests that protein thermostability is achieved as a result of conformational changes in existing proteins and synthesis of proteins more resistant to low-temperature stress.

Chen and Li (1982) have proposed a hypothesis for the hardiness of potato that involves a decrease in osmotic potential as a result of sugar accumulation, an increase in abscissic acid concentration triggered by the lowering osmotic potential, and an increase in protein synthesis caused by the increase in abscissic acid. How broadly this hypothesis can be used remains to be determined.

FACTORS AFFECTING COLD HARDENING

There are a number of factors that affect the reaction of plants to cold hardening. For some factors there are explanations of why they have an effect and for others the explanation is still obscure. Typical of discussions of these factors is that of Gusta et al. (1982) for winter wheat. These authors list the factors in the following manner.

1. *Energy source*. Involves both irradiation intensity and irradiation duration (photoperiod). The concentration of sugars and substrates affecting rates of enzymatic reactions and osmotic potential are involved.

2. *Freezing rate*. For those conditions in which freezing is a factor. By the time that $-5°C$ is reached, over 60% of the water is usually frozen extracellularly. The barriers to efflux of water to the extracellular ice are unlikely to be a limiting factor.

3. *Tissue water content*. The lower the tissue water content, the lower the temperature at which plants survive.

4. *Temperature*. The lowest temperature to which the plants are exposed and the length of time of exposure affect killing temperature. Injury increases with length of exposure time to temperatures approaching the killing temperature.

5. *Water deficit*. Lower water content increases hardiness of the tender, susceptible plant parts compared to the already hardened parts.

6. *Salinity*. The presence of salt reduces hardiness and the rate of hardening even though it may cause a reduction in the crown moisture content.

7. *Nutrition*. Nitrogen fertilization decreases hardiness presumably by increasing tenderness and succulence. Smaller cells, lower water content, and thickened cell walls may be associated with hardiness.

DISCUSSION QUESTIONS

1. What are the low temperature sensitive sites in plants?

2. Distinguish between chilling and freezing injury.

3. Consider a winter wheat plant in which the short axis during the winter is several centimeters below the soil surface. Tiller number has doubled, indicating that growth has occurred. What are the factors that prevent temperature of the crown from falling below the killing temperature?

4. Freeze-injured cells may have a large efflux of sugars and ions but the influx of materials like urea and methyl urea that move passively are not effected. Why? Are the membrane proteins altered?

5. What is meant by freezing or frost hardiness? Is hardiness an inherited or an aquired characteristic?

6. As the temperature drops slowly below the freezing point of water one would expect that the water in the plant would freeze. What are some explanations for why water in the plant may not freeze at $0°C$?

7. What are nucleators? Name some.

8. How can one determine if a plant has frozen?

9. Changes in membranes during cold injury are attributed to lipid phase separation. Explain.

10. What are some of the causes of heat injury?

11. What are the factors that enable a plant to tolerate high temperatures for long periods of time?

12. Based on your knowledge of the physiology of temperature stress and the theories of tolerance, what sorts of manipulations can one perform to render plants more tolerant?

REFERENCES

Asahina, E. 1978. Freezing processes and injury in plant cells. In P. H. Li and A. Sakai (Eds.), *Plant Cold Hardiness and Freezing Stress: Mechanisms and Crop Implications*. Vol. 1. Academic, New York. pp. 17–36.

Barram, L. R., and J. Singh. 1982. Leucine transport in cells isolated from cold-hardened and nonhardened winter rye. *Plant Physiol.* 69:793–797.

Bernstam, V. 1978. Heat effects on protein synthesis. *Annu. Rev. Plant Physiol.* 29:25–46.

Berry, J., and O. Bjorkman. 1980. Photosynthetic response and adaptation to temperature in higher plants. *Annu. Rev. Plant Physiol.* 31:491–543.

Bjorkman, O., M. R. Badger, and P. A. Armond. 1980. Response and adaptation of photosynthesis to high temperatures. In N. C. Turner and P. J. Kramer (Eds.), *Adaptation of Plants to Water and High Temperature Stress*. Wiley-Interscience, New York, pp. 233–249.

Chen, H.-H., and P. H. Li. 1982. Potato cold acclimation. In P. H. Li and A. Sakai (Eds.), *Plant Cold Hardiness and Freezing Stress: Mechanisms and Crop Implications*, Vol. 2. Academic, New York, pp. 5–22.

Davidson, D., and J. Simon. 1981. Thermal adaption and acclimation of ecotypic populations of *Spirodela polyrhiza* (Lemnaceae): Thermostability and apparent activation energy of NAD malate dehydrogenase. *Can. J. Bot.* 59:1061–1068.

Fowler, D. B., L. V. Gusta, and N. J. Tyler. 1981. Selection for winter hardiness in wheat. III. Screening methods. *Crop Sci.* 21:896–901.

Graham, D., and B. D. Patterson. 1982. Responses of plants to low, nonfreezing temperatures: Proteins, metabolism, and acclimation. *Annu. Rev. Plant Physiol.* 33:347–372.

Gullord, M., C. R. Olien, and E. H. Everson. 1975. Evaluation of freezing hardiness in winter wheat. *Crop Sci.* 15:153–157.

Gusta, L. V., D. B. Fowler, and N. J. Tyler. 1982. Factors influencing hardening and survival in winter wheat. In P. A. Li and A. Sakai (Eds.), *Plant Cold Hardiness and Freezing Stress: Mechanisms and Crop Implications*. Vol. 2. Academic, New York, pp. 23–40.

Guy, C. L., and J. V. Carter. 1982. Effect of low temperature on the glutathione status of plant cells. In P. H. Li and A. Sakai (Eds.), *Plant Cold Hardiness and Freezing Stress: Mechanisms and Crop Implications*, Vol. 2. Academic, New York, pp. 169–179.

Hatano, S., T. Kabata, and H. Sadakame. 1981. Transition of lipid synthesis from chloroplasts to a cytoplasmic system during hardening in *Chlorella ellipsoidea*. *Plant Physiol*. 67:216–220.

Hume, D. J., and A. K. H. Jackson. 1981. Frost tolerance in soybean. *Crop Sci*. 21:689–692.

Huner, N. P. A., W. G. Hopkins, B. Elfman, and D. B. Hayden. 1982. Influence of growth at cold-hardening temperature on protein structure and function. In P. H. Li and A. Sakai (Eds.), *Plant Cold Hardiness and Freezing Stress: Mechanisms and Crop Implications*, Vol. 2. Academic, New York, pp. 129–144.

Ilker, R., R. W. Breidenbach, and J. M. Lyons. 1979. Sequence of ultrastructural changes in tomato cotyledons during short periods of chilling. In J. M. Lyons, D. Graham, and J. K. Raison (Eds.), *Low Temperature Stress in Crop Plants— The Role of the Membrane*. Academic, New York, pp. 97–113.

Jordam, S. L., L. S. Jordam, and C. M. Jordam. 1982. Effects of freezing to −196°C and thawing on *Cetaria lutescens* seeds. *Cryobiology* 19: 435–442.

Kramer, P. J. 1969. *Plant and Soil Water Relationships: A Modern Synthesis*. McGraw-Hill, New York.

Krause, G. H., R. J. Klosson, and U. Troster. 1982. On the mechanism of freezing injury and cold acclimation of spinach leaves. In P. H. Li and A. Sakai (Eds.), *Plant Cold Hardiness and Freezing Stress: Mechanisms and Crop Implications*, Vol. 2. Academic, New York, pp. 55–75.

Levitt, J. 1980. *Responses of Plants to Environmental Stresses*. Vol. 1. *Chilling, Freezing and High Temperature Stresses*. Academic, New York.

Li, P. H., and A. Sakai (Eds.), 1982. *Plant Cold Hardiness and Freezing Stress: Mechanisms and Crop Implications*. Vol. 2. Academic, New York.

Lyons, J. M., D. Graham, and J. K. Raison (Eds.), 1979. *Low Temperature Stress in Crop Plants: The Role of the Membrane*. Academic, New York.

McDaniel, R. G. 1982. The physiology of temperature effects on plants. In M. N. Christiansen and C. F. Lewis (Eds.), *Breeding Plants for Less Favorable Environments*. Wiley, New York, pp. 13–45.

Marshall, H. G. 1982. Breeding for tolerance to heat and cold. In M. N. Christiansen and C. F. Lewis (Eds.), *Breeding Plants for Less Favorable Environments*. Wiley, New York, pp. 47–70.

Morris, G. J., and J. J. McGrath. 1981. The response of multilamellar liposomes to freezing and thawing. *Cryobiology* 18:390-398.

Olien, C. R. 1978. Analyses of freezing stresses and plant response. In P. H. Li and A. Sakai (Eds.), *Plant Cold Hardiness and Freezing Stress: Mechanisms and Crop Implications*, Vol. 1. Academic, New York, pp. 37-48.

Olien, C. R., and M. N. Smith. 1981. *Analysis and Improvement of Cold Hardiness*. CRC Press, Cleveland.

O'Neill, S. D., D. A. Priestly, and B. F. Chabat. 1982. Temperature and aging effects on leaf membranes of a cold hardy perennial, *Fragaria virginiana*. *Plant Physiol.* 68:1409-1415.

Palta, J. P., and P. H. Li. 1978. Cell membrane properties in relation to freezing injury. In P. H. Li and A. Sakai (Eds.), *Plant Cold Hardiness and Freezing Stress: Mechanisms of Crop Implication*, Vol. 1. Academic, New York, pp. 93-115.

Pearce, R. S. 1982. Lipids, amino acids, sugars, hardiness, and growth of *Festuca arundinacea*. *Phytochemistry* 21:833-837.

Raison, J. K., J. A. Berry, P. A. Armond, and C. S. Pike. 1980. Membrane properties in relation to the adaptation of plants to high and low temperature stress. In N. C. Turner and P. J. Kramer (Eds.), *Adaptation of Plants to Water and High Temperature Stress*. Wiley, New York, pp. 261-273.

Ritkin, A., C. Gitler, and D. Atsmon. 1981. Chilling injury in cotton (*Gossypium hirsutum* L): Light requirement for the reduction of injury and for the protective effect of abscisic acid. *Plant Cell Physiol.* 22:453-460.

Simonovitch, D. 1981. Common and disparate elements in the processes of adaption of herbaceous and woody plants to freezing—A perspective. *Cryobiology* 18:166-185.

Singh, J. 1981. Isolation and freezing tolerances of mesophyll cells from cold hardened and non hardened winter rye. *Plant Physiol.* 67:906-909.

Steponkus, P. L. 1978. Cold hardiness and freezing injury of agronomic crops. *Adv. in Agron.* 30:481-518.

Steponkus, P. L., and S. C. Wiest. 1979. Freeze-thaw induced lesions in the plasma membrane. In J. M. Lyons, D. Graham, and J. K. Raison (Eds.), *Low Temperature Stress in Crop Plants: The Role of the Membrane*. Academic, New York, pp. 231-254.

Stout, D. G. 1981. Dehydration strain avoidance and tolerance in plant cold hardiness. *J. Theor. Biol.* 88:512-521.

Sullivan, C. Y. 1972. Mechanisms of heat and drought resistance in grain sorghum and methods of measurement. In P. G. Rao and L. R. House (Eds.), *Sorghum in the Seventies*. Oxford and India, New Delhi, pp. 247-264.

Teeri, J. A. 1980. Adaptation of kinetic properties of enzymes to temperature variabiliby. In N. C. Turner and P. J. Kramer (Eds.), *Adaptation of Plants to Water and High Temperature Stress*. Wiley-Interscience, New York, pp. 251-260.

Toivio-Kinnucan, M. A., H. Chen, P. H. Li, and C. Stushnoff. 1981. Plasma membrane alterations in callus tissue of tuber-bearing *Solanum* species during cold acclimation. *Plant Physiol*. 67:478-483.

Trunova, T. I. 1982. Mechanism of winter wheat hardening at low temperature. In P. H. Li and A. Sakai (Eds.), *Plant Cold Hardiness and Freezing Stress: Mechanisms and Crop Implications*, Vol. 2. Academic, New York, pp. 41-54.

Williams, R. J., C. Willemont, and H. J. Hope. 1981. The relationship between cell injury and osmotic volume reduction. IV. The behavior of hardy wheat membrane lipids in monolayer. *Cryobiology* 18:146-154.

Wolfe, J., and P. L. Steponkus. 1983. Mechanical properties of the plasma membrane of isolated plant protoplasts: Mechanism of hyperosmotic and extracellular freezing injury. *Plant Physiol*. 71:276-285.

Wright, L. C., and J. K. Raison. 1981. Correlation between changes in mitochondrial membranes of artichoke tubers and their hardening and dormancy. *Plant Physiol*. 68:919-923.

Yelenosky, G., and C. L. Guy. 1982. Seasonal variations in physiological factors implicated in cold hardiness of citrus trees. In P. H. Li and A. Sakai (Eds.), *Plant Cold Hardiness and Freezing Stress: Mechanisms and Crop Implications*, Vol 2. Academic, New York, pp. 561-573.

5

NUTRIENT STRESS

Nutrient stresses take the form of toxicity or deficiency and have attracted the attention of plant physiologists and crop production scientists for many years. The history of the development of the concepts of mineral nutrient requirements and the methods used to determine essentiality of some of the elements, the criteria that have been used to designate essentiality, and studies of the factors affecting availability, translocation, and metabolic roles have had the attention of many investigators and yet there are many unanswered questions.

Approximately one fourth of the earth's soils are considered to produce some kind of mineral stress (Dudal, 1976). This estimate does not include the effects of pollution from human activities on the environment in which plants grow; nor does it include the depletion effects of intensive agriculture on the mineral element reserves in the soil.

Approximately 10% of the plant dry weight is composed of minerals or about 1.5% of the fresh weight. Currently 15 to 17 elements are considered to be essential for plant growth and reproduction. Of these, the physiological role in metabolism of only a few is known in detail. Most plant analyses are made on ground and ashed material that tells little of the function of the elements. There are numerous published tables of nutrient content of many agronomic and food crops, which, when coupled with knowledge of

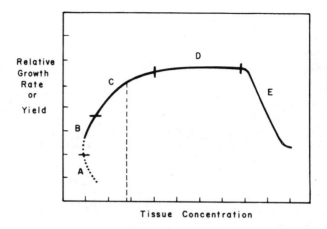

Figure 5.1. Diagram of the general effects of available nutrient concentration in the rooting medium or, of nutrient tissue concentration, on growth. (A) Extreme deficiency range; (B) Severe deficiency range; (C) Moderate deficiency range; (D) Luxury range; (E) Toxic range. The vertical dashed line indicates the critical concentration. One convention has the critical concentration at the point where there is a 10% reduction of growth.

the soils and the past history of the land on which the crops are growing, can be used to determine whether there is a deficiency or toxicity of an element. Interpretation of such information is more an art than a science. The range of concentrations that may be found is expressed in Figure 5.1. Arbitrarily determined ranges are severe deficiency, moderate deficiency, luxury consumption, and toxic. The concept of critical concentration range has been described as that range of concentrations that occurs between deficiency and luxury consumption. Such information is useful in measuring the degree of stress a plant may be experiencing and whether it is in the deficient or toxic range of concentrations.

CONDITIONS CAUSING NUTRIENT STRESS

Deficiency of an element results from a number of conditions among which are the amounts and concentrations present in the soil, the form in which they exist, the processes by which they become available to the plant, the content of the soil solution, and the soil pH. The form of the nutrient avail-

TABLE 5.1   Form in Which the Essential Elements Are Available to Plants

Essential Element	Form Available to Plants	Source	Concentration in Dry Tissue	Concentration in Soils
		Micronutrients		
Mo	MoO_4^-	Soil solids	0.1 ppm	2 ppm
Cu	Cu^+	Soil solids	6	50
	Cu^{2+}			
Zn	Zn^{2+}	Soil solids	20	100
Mn	Mn^{2+}	Soil solids	50	2,500
Fe	Fe^{2+}	Soil solids	100	250,000
	Fe^{3+}			
B	BO_3	Soil solids	20	100
	B_4O_7			
		Macronutrients		
S	SO_4^-	Soil solids	0.1 %	0.04 %
P	H_2PO_4	Soil solids	0.2	0.05
Mg	Mg^{2+}	Soil solids	0.2	0.03
Ca	Ca^{2+}	Soil solids	0.5	0.43
K	K^+	Soil solids	1.0	0.17

able to plants is given in Table 5.1 as are the relative amounts in soils and the relative concentrations found in plants. Nutrient concentrations in plants vary with age, plant part, species, and the rooting medium or soil type. The amounts of nutrients in soil vary widely and depend upon the parent material from which the soil is derived and the processes in cycling the elements in natural or agricultural ecosystems. A discussion of these factors and the chemical reactions in soil that account for the various forms of the elements available can be found in elementary soils texts.

Nutrient absorption from soil involves movement from soil to the root, from the exterior of the root to the interior, and the translocation within the plant. Movement through the soil to the root may be by diffusion in the soil solution, diffusion from soil particle to the root directly, diffusion from one particle to another, replenishment from the parent material or the adsorption surfaces to the soil solution, or by mass flow in the soil solution by capillary action or gravitational flow. Growth of roots through the soil also brings them into contact with fresh areas of soil and ions become available by this process of interception. Ability of plant roots to exude into the rhizosphere materials that affect the solubility and form of nutrients is also well known. The role of exudates is discussed later.

Absorption of nutrients by plant roots decreases the concentration of ions in the vicinity of the root and sets up concentration gradients toward the root. The surface area of the root system relative to its volume thus becomes crucial in the absorption process. Root hairs and mycorrhizal fungal hyphae have a role in the absorption of ions by increasing the absorbing surface.

In general, if a large amount of an ion is bonded to exchange sites, then a large amount of that ion can be obtained by root interception provided there is much root-soil contact. If much of the ion is in solution and replenishment is fast, mass flow can supply much of the nutrient to the root. But, if the solution concentration is low and replenishment is slow, diffusion is the main mechanism moving the ion to the root surface.

Interference with any of the processes involved in changing the availability of ions or of their absorption by the plant roots may result in stress. In addition, some factors that affect the intensity and site of expression of mineral deficiency or toxicity in plants need to be understood. Some of the ions are mobile within the plant and some are immobile. That is, some ions are capable of moving from older leaves, for example, to younger growing parts of the plant. Other ions are fixed and stay where they are incorporated into the structure of the plant or its metabolites. Still other ions accumulate to high concentrations in vacuoles of leaves (aluminum, for example). The fluxes that occur are the result of translocation either in the xylem or the phloem along concentration or pressure gradients and depend on the form in which the ion is translocated. Movement in the xylem shows little or no difference in rate of translocation of the various ions and is controlled by transpiration rates and the pressure gradients created by differences in water potential. In the phloem there are distinct rate differences for translocation of the various ions and such differences may produce difficiencies in some parts of the plant. Elements such as calcium, boron, and iron often become deficient in the younger parts of the plant because of little or no retranslocation and poor phloem mobility. Similarly, if such ions are applied foliarly, there is little basipetal movement. Toxic concentrations of ions often appear in older tissues of the plants as a result of accretion with time.

DEFICIENCY CAUSES AND SYMPTOMS

A symptom is any perceptible change in known structure, appearance, or function. Such things as leaf yellowing (chlorosis), death (necrosis), lesions, malformations, malfunctions, or reduced growth and yield are all considered to be symptoms and occur as a result of nutrient deficiency. However, confusion occurs when one realizes that similar symptoms can be produced by deficiencies, toxic levels of nutrients, pathogenic organisms, air pollutants, and pesticides, and symptoms may vary with plant species or variety. Combinations of two or more causes of symptoms may result in a situation that is difficult to diagnose. An example of foliar deficiency symptoms and how they may be used to determine which deficiency one is observing is presented in Table 5.2. The information upon which the key is based was derived from tobacco plants growing in sand culture in the greenhouse. There are keys for other species and varieties of specific crops.

Yield or growth rate is usually impaired before other symptoms of deficiency occur so that some other indications of at least the potential for a deficiency must be considered in any diagnosis procedures. What this suggests is that diagnosis should be left to the experts who are trained to look at all possible causes and have available the necessary testing facilities.

TABLE 5.2   Key to Plant Nutrient Deficiency Symptoms

Symptom	Deficient Element
A. Parasitic and virus diseases disseminated by bacteria, fungi, or virus entities (excluded from present discussion).	
A. Nonparasitic troubles; never infectious; caused by element deficiencies.	
B. Older or lower leaves of plant mostly affected; effects localized or generalized.	
C. Effects mostly generalized over whole plant; more or less drying or firing of lower leaves; plant light or dark green.	
D. Plant light green; lower leaves yellow, drying to light brown color; stalks short and slender if element is deficient in later stages of growth................	NITROGEN
D. Plant dark green, often developing red and purple colors; lower leaves sometimes yellow, drying to greenish brown or black; stalks short and slender if element is deficient in later stages of growth........	PHOSPHORUS
C. Effects mostly localized; mottling or chlorosis with or without spots of dead tissue on lower leaves; little or no drying up of lower leaves.	

TABLE 5.2 *(Continued)*

Symptom	Deficient Element
D. Lower leaves mottled or chlorotic, with or without dead spots; leaf margins sometimes tucked or cupped upward or downward.	
E. Mottled or chlorotic leaves typically, may redden, as with cotton; sometimes with dead spots; tips and margins turned or cupped upward; stalks.slender . .	MAGNESIUM
E. Mottled or chlorotic leaves with large or small spots of dead tissue.	
F. Spots of dead tissue small, usually at tips and between veins, more marked at margins of leaves; stalks slender	POTASSIUM
F. Spots generalized; rapidly enlarging, generally involving areas between veins and eventually involving secondary and even primary veins; leaves thick; stalks with shortened internodes. . .	ZINC
B. Newer or bud leaves affected; symptoms localized.	
C. Terminal bud dies, following appearance of distortions at tips or bases of young leaves.	
D. Young leaves of terminal bud at first typically hooked, finally dying back at tips and margins, so that later growth is characterized by a cut-out appearance at these points; stalk finally dies at terminal bud	CALCIUM
D. Young leaves of terminal bud becoming light green at bases, with final break down here; in later growth, leaves become twisted; stalk finally dies back at terminal bud	BORON
C. Terminal bud commonly remains alive; wilting or chlorosis of younger or bud leaves with or without spots of dead tissue; veins light or dark green.	
D. Young leaves permanently wilted (wither-tip effect) without spotting or marked chlorosis; twig or stalk just below tip and seedhead often unable to stand erect in later stages when shortage is acute	COPPER
D. Young leaves not wilted; chlorosis present with or without spots of dead tissue scattered over the leaf.	
E. Spots of dead tissue scattered over the leaf; smallest veins tend to remain green, producing a checkered or reticulated effect	MANGANESE
E. Dead spots not commonly present; chlorosis may or may not involve veins, making them light or dark green in color.	
F. Young leaves with veins and tissue between veins light green in color	SULFUR
F. Young leaves chlorotic, principal veins typically green; stalks short and slender.	IRON

Source. J. E. McMurtrey, Jr. 1948. *Diagnostic Techniques for Soils and Crops.* American Potash Institute.

DEFICIENCY AND TOXICITY CAUSES AND EFFECTS

To illustrate the principles of cause and effect we have chosen to use three examples. A proper cataloging of all the essential elements, their role in metabolism, and the alteration of metabolism when they are deficient is contained in monographs and other works that are listed at the end of the chapter. The three examples we have chosen are copper, manganese, and iron, all of which are transition metals and micrometabolic elements.

Copper

Copper exists in the two oxidation states, cuprous and cupric. Even weak reducing agents can reduce the cupric ion and so the cuprous form is the more stable. The cuprous ion binds preferentially to oxygen and carbon monoxide and oxidizes them. There is also a strong binding to organic ligands such as peptides. Because of a strong competition for ligands, copper becomes toxic at the relatively low levels of 3 to 10 ppm. For this reason tolerant plants may have many ligands as amino groups on proteins that chelate copper and prevent toxicity.

Because copper acts as a catalyst in oxidation–reduction reactions in the plant, its deficiency affects the energy flow in respiration and photosynthesis. Copper alone is a catalyst but when complexed with a protein it is a 1000-fold more active. The chelated forms of copper occur in ascorbic acid oxidase and polyphenol oxidase. Copper also occurs in the photosynthetic electron transport system mediated by plastocyanin and in the terminal oxidation by cytochrome oxidase in respiration.

Deficiency of copper reduces the yield and quality of fruit and grain and under severe deficiency may also reduce vegetative growth. The most apparent effects of copper deficiency in plants are related to lignification. Under severe deficiency the plants show reduced lignification and thickening of cell walls of xylem elements, fibers and epidermal cells (Schutte and Mathews, 1968). Under more severe conditions the phloem may be damaged. Such responses may be the result of the role of copper in the polyphenol oxidase reactions (Judel, 1972). The lack of lignification can be a specific indicator of copper deficiency (Bussler, 1981).

Manganese

In contrast to copper, manganese has the lowest stability constant in chelates of any of the transition elements. In the oxidized state the manganese atom loses the 4s electron pairs so that a single electron is left in each of five 3d orbitals. The octahedral coordination is of a covalent nature, which makes manganese involved in protein structure rather than in redox reactions. Tightly bound manganese is involved in splitting water in photosystem II but the structure and function are not known. Manganese is somewhat mobile in the plant and the deficiency symptoms first appear in the lower or older leaves. The symptoms are characterized by a color lightening of the interveinal tissue in broadleaved plants and long necrotic streaking in the interveinal tissue of cereal and grass leaves.

Iron

Iron is functional in the ferrous state and is bound in stable complexes by chelating nitrogen donors where it may be oxidized to the ferric state. The most studied roles of iron are in the oxidation–reduction reactions of iron proteins such as cytochrome, hemes, ferrichrome, and leghemoglobin. Some of the familiar exzymes in which iron is the active site are cytochrome oxidase, catalase, peroxidase, and nitrogenase.

The ferric ion is the most mobile and because of its relatively small radius is a strong Lewis acid and may be precipitated by phosphates or hydroxides. There is a strong interaction between iron and phosphorus in the metabolism of plants that sometimes results in phosphorus deficiency when there are relatively high concentrations of iron in the environment. Tissue analysis as ordinarily performed does not distinguish between functional iron and that inactivated by precipitation or complexing. Active iron is that which is extractable by normal hydrochloric acid from dried leaf powder. Under some conditions of rapid growth there are enough ligands produced to complex iron to the extent that it may become deficient for metabolic functions. In iron deficient leaves there are large amounts of free amino acids that disappear upon the addition of iron (Dekock, 1981).

Factors that result in iron induced chlorosis of plants may be grouped into those that affect the availability of iron in the soil, those that affect the ability of the roots to absorb iron, and those that affect the translocation of iron to its sites of action. The effects of some of the environmental factors that produce iron stress are clear and logical but others are obscure. The following is a list of factors that cause iron deficiency symptoms:

1. Low iron supply
2. High calcium carbonate (lime induced chlorosis)
3. Over irrigation, soil saturation
4. High phosphate concentrations
5. High concentrations of heavy metals such as manganese and copper
6. Either high or low temperatures
7. High light intensities
8. High levels of nitrate
9. Some unbalanced cation ratios
10. Poor aeration
11. Some additions of organic matter to soil
12. Virus infections

A deficiency of iron appears first in the youngest leaves as an interveinal chlorosis. Later, entire leaf blades may become yellow or white. The leaf veins are the last to lose chlorophyll. Does the loss of chlorophyll result from a decrease in the rate of synthesis or from an increase in the rate of destruction, or both? Just how iron affects chlorophyll synthesis is unknown. It may play a role in the synthesis of some precursor, such as the porphyrin rings, in the metabolic pathway.

Translocation of iron to the leaves occurs in the form of citrate where it may be stored in granules of phytoferritin. Once deposited in the leaves iron is relatively immobile.

PLANT ANALYSIS AS A DIAGNOSTIC TOOL

Plant analysis, when used as a means of assessing the nutritional requirements of plants, may be envisaged as a study of the relationship of nutrient content of the plant to the growth of the plant. From a physiological point of view, progress in plant analysis can be made only when the function of each nutrient is understood. Although much progress has been made in understanding the role of nutrients in plant metabolism, there is much research to be done. Over the years a trial and error method has been used to appraise the nutrient requirements of crop plants.

Analyzing plant material to estimate the nutrient content of soil has proven unsatisfactory because plants can accumulate high concentrations of some elements and exclude others. If the levels in plant tissues that are sufficient for growth are known, various tests can be used to determine the levels in the plant. Concentrations that are deficient, sufficient, or toxic for the micronutrients are presented in Table 5.3. This is a general table compiled from tissue analyses from many plants over several years, but to be of practical value such tables are needed for specific crops on specific sites.

Although there are fresh tissue analyses, their use is limited. The usual methods employ dried and ground tissue that is oxidized to remove the organic material followed by the measurement of the concentration of the various elements present in the resulting ash. Two concepts have been developed as a result of these types of analyses: (1) the concept of critical nutrient concentration and (2) the concept of nutrient balance.

The critical nutrient concentration with respect to growth may be defined as that nutrient concentration that is just deficient for maximum growth, or that is just adequate for maximum growth, or that separates the zone of deficiency from the zone of adequacy. As determined experimentally, the critical concentration is not a point but a narrow range of nutrient concentrations above which the plant is amply supplied with nutrients

TABLE 5.3   Micrograms of Micronutrient per Gram Dry Weight (ppm) in Mature Leaves

Micronutrient	Deficiency Range	Sufficiency Range	Excessive or Toxic
B	<15	20–100	200 +
Cu	4	5–20	20 +
Fe	50	50–250	Not known
Mn	20	20–500	500 +
Mo	0.1	0.5	Not known
Zn	20	25–150	400 +

Source. J. B. Jones, Jr. 1972. Plant tissue analysis for micronutrients. In J. J. Mortvedt, P. M. Giordana, and W. L. Lindsey (Eds.), *Micronutrients in Agriculture.* Used with permission from the Soil Science Society of America, Madison, WI.

and below which the plant is deficient. Ideally, the critical concentration of a nutrient just deficient for maximum growth should be determined by observing a tissue, a single plant cell, or even a specific physiological reaction. However, since none of the suggested procedures are convenient or even feasible at this time, alternative methods must be used. The most direct method is through the use of solution, sand, or soil cultures, or through field experiments in which a series of plants in pots or field plots are supplied with an increasing amount of the nutrient under study, while all other nutrients are maintained at adequate levels. The plant leaves are sampled when some members of a series are deficient and others are still amply supplied with the nutrients being studied. A critical nutrient concentratrion may be estimated less accurately by analyzing material from plants showing characteristic deficiency symptoms. The latter approach is feasible only when symptoms can be recognized readily and when they manifest themselves in a constant manner.

In contrast, the nutrient balance concept may appear to be an opposing point of view but should be considered as an extension of the critical nutrient concept. In the critical nutrient concept, the assumption is made that there is a relatively narrow range of concentrations for a given nutrient at which growth begins to decrease. It is further assumed that wide variations, short of excesses in other nutrients, have no large influence on the growth of the plant and that a major increase in growth is obtained only through the direct or indirect addition of the deficient nutrient. In the nutrient balance concept, the proportion of nutrients in the dried tissue is considered, as well as the actual concentration. While it appears logical to consider all essential elements as having a specific balance in the proper concentrations for maximum growth, the practical difficulty of demonstrating the uniqueness of the given balance is not easily attained experimentally. Until more is known of the functions of the nutrient elements within the plant, it appears to be more expedient to study the relationships of the nutrients to plant growth than to study them in respect to physiological processes.

There are a number of factors that affect the critical concentration. Concentrations in young tissues may be diluted as growth proceeds (Table 5.4), but for other nutrients, the concentration may be low in young tissues and increase as growth continues because of the gradients toward the young tissues and the differential mobility of the various elements (Table 5.5). The accumulation of dry weight in organic matter of the plant dilutes

TABLE 5.4 Micrograms of Micronutrients per Gram Dry Weight in Soybean Leaves, Stems, and Pods at Various Periods During the Growth Cycle

Micronutrient	Leaves 10 Days after Emergence	Leaves at Pod Set	At Maturity		
			Stems	Pods	Leaves
B	50	60	23	38	62
Cu	12	15	7	12	13
Fe	420	550	75	75	400
Mn	130	120	20	30	150
Zn	38	45	19	40	45

Source. J. B. Jones, Jr., 1972. Plant tissue analysis for micronutrients. In J. J. Mortvedt, P. M. Giordana, and W. L. Lindsey (Eds.), *Micronutrients in Agriculture*. Used with permission from the Soil Science Society of America, Madison, WI.

TABLE 5.5 Trends in Leaf or Fruit Concentrations of Minerals in Relation to Age of Tissue of Various Plants

Plant	Decrease with Age	Increase with Age
Apple	N, P, K	Ca, Mg
Blueberry	P	Ca, Mg
Citrus	N, P, K, Cu, Zn	Ca, Mg, Mn, Fe, Al, B
Citrus fruit	N, P, K, Mg	Ca
Fig	N, P, K	Ca, Mg
Peach	N, P, K, Cu, Zn	Ca, Mg, Mn, Fe, Al, B
Pine	K	Ca
Vegetables	N, P, K	Ca

Source. P. F. Smith. 1962. Mineral analysis of plant tissues. *Annu. Rev. Plant Physiol.* 13:81–108. Used with permission from Annual Reviews Inc., Palo Alto, CA.

all the elements if concentration is expressed as milligrams per gram or percent dry weight.

Interactions of one nutrient with another also affect the concentration. Increasing the supply of one nutrient may depress leaf concentrations of other nutrients either by increasing growth or interfering with absorption or translocation (Table 5.6). Continued growth of new plant parts may alter the concentrations in older leaves with subsequent replacement.

The formation and development of fruits may create gradients that cause nutrients to go directly to the fruits and bypass the vegetative parts of the plant. Fruits compete favorably with leaves for translocated nitrogen, for example. Fruits also have a high requirement for potassium. There are

TABLE 5.6 General Effect of an Applied Element on the Mineral Composition of Citrus Leaves[a]

Element Added	Elements Measured in Leaves									Total No.
	N	P	K	Ca	Mg	Cu	Zn	Mn	B	
N	+	−	−	+	+	−	−	−	0	8
P	0	+	−	+	−	−	+	+	0	7
K	−	0	+	−	−	0	0	0	−	5
Ca	0	0	−	+	0	?	?	?	?	2
Mg	0	0	−	−	+	−	+	+	0	6
Cu	0	0	+	0	0	+	−	−	0	4
Zn	0	0	+	−	−	−	+	−	0	6
Mn	0	0	+	−	−	0	0	−	+	6
B	0	−	+	−	−	0	0	−	+	4
Total No.	2	3	9	7	7	6	5	7	2	48

Source. P. F. Smith. 1962. Mineral analysis of plant tissues. *Annu. Rev. Plant Physiol.* 13:81–108. Used with permission from Annual Reviews Inc., Palo Alto, CA.

[a] +, increase; −, decrease; 0, no change; Total No., number of elements that changed in concentration.

seasonal effects in the concentrations in leaves, either because of age differences in the tissues or because of phenological differences.

NUTRIENT AND METAL TOXICITY

The range of concentrations of the essential nutrients between deficiency and toxicity is often small and depends on a variety of factors in the environment that affect availability and growth of the plant. The transition metals (micrometabolic nutrients) are most often implicated in toxicity reactions. Physiological studies of toxicity are difficult because the active concentration of a metal or its salts is unknown as a result of chelation and the effects of pH on solubility. Studies using isolated organelles have been useful but the concentration of metals has often been unrealistically high and the species of the chemical and ion activity have been unknown. Toxicity implies a sensitivity of an "essential to life" process or processes. The site of toxic action is difficult to find and there is insufficient information on the processes involved.

Toxic levels of zinc, copper, and nickel occur frequently in soil. Toxic concentrations of lead, cobalt, beryllium, and cadmium can occur but only under unusual conditions. Lead and cadmium are of particular interest because plant accumulation introduces them into the food chain. Aluminum and manganese toxicities are the most studied and the information about them can serve as a basis for some concepts and principles that may apply to the others (Foy et al., 1978).

ALUMINUM TOXICITY

The appearance of plants suffering from aluminum toxicity gives clues to the physiological effects. For some plants the effects resemble those of phosphorus deficiency, suggesting that there may be a cause and effect relationship between aluminum toxicity and phosphorus uptake. The symptoms are stunting, small dark green leaves, and late maturity with a purpling of the leaves and stems. Often there is a yellowing and death of the leaf tips. On other plants there is an induced calcium deficiency that results in curling or rolling of young leaves, and a collapse of growing points or petioles. Roots are injured and become brittle and stubby with a spatulate shape. Root tips become thickened and turn brown and there are many stubby lateral roots with no fine branching, which gives rise to a coralloid appearance. Such roots are predisposed to fungal infection and the absorption processes are inefficient.

Little information is available about the effects of aluminum toxicity on the physiological processes. Interference with cell division should be suspected because of the effects on root tips. The cells become binucleate and if aluminum penetrates to the nuclei, it is precipitated there by the phosphorus containing compounds. There is an interference with the enzymes governing the deposition of cell wall polysaccharides and, because of the cross linking of aluminum in the pectins, the walls become rigid. Cross linkages within proteins also occur. Probably because of the change in pattern of root growth, there is interference with the uptake and transport of other essential elements such as calcium, magnesium, phorphorus, and potassium.

The aluminum–phosphorus interaction is well known and has been studied extensively. An aluminum–phosphorus precipitate may occur extracellularly along the root surface or in the cell wall of root caps and the first few cortical cells of the root. Aluminum is rarely observed to be translocated to any appreciable degree in most temperate aluminum sensitive plants. It usually enters roots in areas of the emergence of lateral roots or through meristems as they differentiate. But in some plants, aluminum is readily removed from the roots and accumulates in the leaves to high concentrations. In *Symplococcus spicata* concentrations as high as 72,300 ppm have been found. Other species are known with concentrations in the 2000- to 3000-ppm range. Most plants, however, accumulate concentrations only in the 200-ppm range, which is in the range of concentrations of the micrometabolic nutrients.

In general, there are three groups of plants: (1) the aluminum excluders, which include cultivars of wheat, barley, and soybeans; (2) the entrappers, which trap aluminum in the roots preventing it from reaching the leaves, include azalea, cranberry, rice, rye, and alfalfa clones; (3) the accumulators in which aluminum is translocated to the leaves where it reaches high concentrations. This group includes tea, some grasses, pine trees, and mangroves. Although this may be a good way to group plants in their reaction to aluminum, it tells little about the mechanisms that render some plants tolerant and others intolerant to aluminum.

Foy et al. (1978) discuss some of the physiological factors associated with differential tolerance. The exact role of these factors in tolerance is not clear and different biochemical pathways may be involved in different species with different genotypes. The pH in the root zone affects solubility, and the ability of the plant to change the pH, particularly in the rhizosphere, may affect the tolerance mechanism. Such pH changes may also be related to the source of nitrogen in the rooting medium. The preferential uptake of nitrate over ammonium sources of nitrogen will cause a rise in pH. Release of organic acids or of hydrogen ions from root surfaces will also affect the pH of the rhizosphere. Detoxification of aluminum by chelation with organic acids or other complex organic molecules in cells of tolerant plants is also a possibility.

MANGANESE TOXICITY

Manganese is an example of one of the essential micrometabolic elements in which the range in concentrations between deficiency and toxicity may be quite narrow. The element accumulates in the shoots of plants and the injury is in proportion to the amount accumulated. Symptoms include a marginal chlorosis and necrosis of the leaves, leaf puckering, and formation of necrotic spots. The symptoms may be produced without a reduction of the vegetative growth. Severe injury to the shoot may later result in the roots turning brown. The physiological effects vary from species to species and have been described using a variety of species. Manganese toxicity has been associated with the destruction of auxin in cotton (Morgan et al., 1966; 1976) and with an amino acid imbalance in potato (Robinson and Hodgson, 1961). As with aluminum, tolerance of high manganese concentrations in certain plants has been attributed to reduced absorption, reduced translocation to the shoots, and less sensitivity of the tissues to manganese. Internal tolerance of high manganese concentrations varies widely among plant species and may be from 380 to 1600 ppm. The interaction between iron and manganese may result in iron deficiency and chelated forms of iron may decrease the absorption of manganese. Hence, it is useful to know that foliar sprays of iron may counteract the toxicity of manganese.

COPPER TOXICITY

Copper is another of the transition elements, a group characterized by the possession of a partly filled set of d orbitals. A wide variety of organic ligands are known to form chelates with copper and the usual number of monodentate ligands per copper atom is four. Because of this affinity for organic ligands, copper is found as the prosthetic group in the group of enzymes having to do with oxidation–reduction reactions in plants (Lepp, 1981).

Copper toxicity generally results in chlorosis and stunting. The chlorosis is attributed to iron deficiency caused by the high concentrations of copper in the tissues. Toxicity is experienced first in the root tips followed by inhibition of the development of lateral roots. Reduction of the root system has implications for the absorption of other nutrients and the subsequent growth inhibition. The effect of copper on roots has not been explained. Some attempts have been made to link the toxicity and reduction of root growth to the effect of copper on inhibition of indole acetic acid oxidase activity (Coombes et al., 1976).

MYCORRHIZAE AS A FACTOR IN STRESS ALLEVIATION

Nutrient influx takes place across the soil-root interface including the complex of materials and organisms in the rhizosphere. Under conditions of limited nutrient supply, mycorrhizae may be formed and cause an increase in nutrient uptake. It has been estimated that more than 80% of the taxa of higher plants have mycorrhizal roots. It is the ectomycorrhizal association that increases mineral nutrient absorption and there are many factors involved in the relationship. The fungal hyphae increase the surface area in contact with the soil and access more volume of soil containing nutrients. The life of mycorrhizal roots is prolonged so that absorption from a soil site occurs over a longer period of time than would occur with nonmycorrhizal roots. Also, mycorrhizal roots absorb nutrients more rapidly per unit area than nonmycorrhizal roots. Because of the formation of a mantle of fungal hyphae over the root surface, it may be protected from invasion by pathogenic organisms. Mycorrhizae may also increase tolerance to toxic levels of essential metal ions in the rooting medium. The explanation is that the formation of precipitates and complexes of metals is greater in mycorrhizal roots than in nonmycorrhizal roots; hence, there is a decrease in foliar concentration of the metals.

Mycorrhizal roots seem to form more abundantly in soils of low fertility and so they are of greater prominence in woody species than in crop species which grow on relatively fertile soil. Foresters have known for some time that trees planted in areas where they have not grown previously need to be inoculated with fungi that form mycorrhizae or the trees will be stunted and lack vigor. The role of the mycorrhizal association in stress situations is an important one and its presence may be the difference between survival of the plant and death.

CHELATION AS A MECHANISM OF TOLERANCE

Chelation is the coordination between a metal and a ligand resulting in the formation of a ring structure. Usually two or more ligands are necessary for the ring structure. Ligands are compounds containing functional groups of atoms that share electrons with metals. They are bases with unshared electron pairs. The metals act as acids and accept electron pairs.

The neutralization reaction between these Lewis bases and Lewis acids is called coordination. The resulting bonding may not be stationary but may shift according to a statistical probability based on the quantum mechanics involved.

The properties of the metal ion that determine the interaction with coordinating ligands include valency, crystal ion radius, polarizability, hydration energy, hydrated ion radius, mass, and coordination number. The properties of the ligand on the other hand are its spatial requirement related to ring size and configuration and the nature of the ligand molecules. In the formation of chelates the arrangement of electrons in orbitals around the nucleus may be altered.

The relationship of chelates to mineral element stress tolerance lies in the structure and stability of the chelate ring, which is a function of the number and size of the rings and the composition of the ligands. Because chelation has an effect on the reactivity of the metal, plants may be protected from excessive concentrations of free ions and may have increased stabilization of structural elements.

Chelation results in the formation of metalloenzymes, which perform vital functions in plant metabolism. Changes in metabolism may come about because of the changes in enzyme composition as a result of stress and the effects on available ligands, metal ion concentrations, and the proteins associated with enzymes. Some metalloenzymes are located in membranes where deficient or toxic concentrations of metal ions may affect permeability. Calcium, for example, reacts with acid phosphatase or acid proteins in the membranes.

GENETICS OF MINERAL ELEMENT STRESS TOLERANCE

The degree of heritability and gene action involved in the efficient utilization of the essential elements has not been intensively investigated. For some crops, the levels of segregation of efficient and inefficient strains were significant enough that selection should be possible in breeding programs (Clark, 1982).

Considerable research has been done on the genetics of metal toxicities. In barley and wheat, for example, tolerance to aluminum has been found to be controlled by a single, major, dominant gene in some cultivars but a more complex situation existed in other cultivars (Clark, 1982). Apparently, there is sufficient evidence that for cereals and other plants genetic improvement for aluminum and manganese tolerance can be achieved.

The technique of tissue culture has been used to produce lines that contain chromatin material from other lines brought about by a transfer manipulation that cannot happen naturally. For example, the copper efficiency factor from rye was transferred to triticale and then to wheat by modern technology.

The genetics of iron efficiency have been investigated extensively. Plants that exude hydrogen ions and reducing compounds from their roots or are capable of reducing iron at the root surfaces can survive on soils with low iron availability, but plants that do not contain these traits may not survive. Iron efficient plants under iron deficiency conditions develop epidermal transfer cells in the subapical region of the root. The transfer cells are involved in the extrusion of protons into the soil, which results in a lower pH. They may also be involved in the production of reducing substances. An abundance of root hairs develops that may contain transfer cells as well. A rapid increase in iron uptake occurs as a result of these changes in root morphology and anatomy (Romheld et al., 1982).

Transfer cells are recognizable by the dense cytoplasm and the accumulation of mitochondria in the region near the outer cell wall. The cells have an active proton pump mechanism. Other characteristics of the transfer cells are described by Landsberg (1982).

Another factor in tolerance is that plants that are tolerant of iron deficiency conditions frequently contain higher concentrations of elements such as phosphorus, calcium, zinc, and manganese than the less tolerant plants.

Because of the importance of overcoming nutrient stress in soils of countries where there is promise of increases in yield, there is an international cooperation in research on ways of bringing about adaptation of important crop plants to the stress.

DISCUSSION QUESTIONS

1. Give three reasons why it is important to understand elemental cycling.

2. On what basis is nutrient concentration of plant tissues usually expressed?

3. What is meant by the terms: critical concentration, luxury consumption, nutrient balance, deficiency, toxicity?

4. What factors affect deficiency levels of elements in plants?

5. What are some possible reasons for the fact that nitrogen deficiency becomes evident in plants before sulfur deficiency even though some of the symptoms are similiar?

6. Outline the various factors that may limit nutrient absorption from the soil by plant roots.

7. What is meant by mobile and immobile elements? What role does mobility play in nutrient stress?

8. What cautions must be considered in the diagnosis of mineral deficiency?

9. How can one determine the critical concentration of an element and what practical use can be made of this knowledge?

10. What role do mychorrhizal roots play in the alleviation of nutrient stress?

11. Why are aluminum, manganese, and copper sometimes toxic?

12. What are the most prominent aspects of chelation and the roles of chelates in physiological processes?

REFERENCES

American Phytopathological Society. Various compendia on plant diseases. 1975 to present.

American Society of Agronomy. 1978. Crop tolerance to suboptimal land conditions. ASA Special Publication Number 32, Madison, WI.

Bjorkman, E. 1970. Forest tree mycorrhiza—The conditions for its formation and the significance for tree growth and afforestation. *Plant Soil* 32:589-610.

Bussler, W. 1981. Microscopical possibilities for the diagnosis of trace element stress in plants. *J. Plant Nutr.* 3:115-125.

Clark, R. B. 1982. Plant response to mineral element toxicity and deficiency. In M. N. Christiansen and C. F. Lewis (Eds.), *Breeding Plants for Less Favorable Environments*. Wiley, New York, pp. 71-142.

Clarkson, D. T., and J. B. Hanson. 1980. The mineral nutrition of higher plants. *Annu. Rev. Plant Physiol.* 31:239-298.

Coombes, A. J., N. W. Lepp, and D. A. Phipps. 1976. Effect of copper on IAA-oxidase activity in root tissue of barley (*Hordeum vulgare* L cv Zephyr). *Z. Pflanzenphysiol.* 80:236-242.

Dekock, P. C. 1981. Iron nutrition under conditions of stress. *J. Plant Nutr.* 3:513-521.

Dudal, R. 1976. Inventory of the major soils of the world with special reference to mineral stress hazards. In M. J. Wright (Ed.), *Plant Adaptation to Mineral Stress in Problem Soils.* Cornell Univ. Agr. Exp. Sta., Ithaca, NY, pp. 3-13.

Foy, C. D., R. L. Chaney, and M. C. White. 1978. The physiology of metal toxicity in plants. *Annu. Rev. Plant Physiol.* 29:511-566.

Gerdeman, J. W. 1974. Mycorrhizae. In E. W. Carson (Ed.), *The Plant Root and Its Environment. University Press of Virginia*, Charlottesville, pp. 205-217.

Hacskaylo, E. 1972. Mycorrhiza: The ultimate in reciprocal parasitism? *Bioscience* 22:577-583.

Harley, J. L. 1969. *The Biology of Mycorrhiza.* 2nd ed. Leonard Hill, London.

Hewitt, E. J., and T. A. Smith. 1975. *Plant Mineral Nutrition.* English Universities Press, London.

Huisingh, D. 1974. Heavy metals: Implications for agriculture. *Annu. Rev. Phytopathol.* 12:375-388.

Jones, J. B., Jr. 1972. Plant tissue analysis for micronutrients. In J. J. Mortvedt, P. M. Giordana, and W. L. Lindsey (Eds.), *Micronutrients in Agriculture.* Soil Sci. Soc. Am., Madison, WI.

Judel, G. K. 1972. Effect of copper and nitrogen deficiency on phenol oxidase activity and content of phenols in leaves of sunflower (*Helianthus annuus*). *Z. Pflanzenernaehr. Bodenkd.* 131:159-170.

Landsberg, E. C. 1982. Transfer cell formation in the root epidermis: A prerequisite for Fe efficiency? *J. Plant Nutr.* 5:415-439.

Lepp, N. W. 1981. Copper. In N. W. Lepp (Ed.), *Effect of Heavy Metal Pollution on Plants.* Applied Science, Liverpool, pp. 111-143.

Lepp, N. W. (Ed.) 1981. *Effect of Heavy Metal Pollution on Plants.* Vol. 1. *Effects of Trace Metals on Plant Function.* Applied Science, Liverpool.

McMurtrey, J. E., Jr. 1948. *Diagnostic Techniques for Soils and Crops.* American Potash Institute, Atlanta.

Marks, G. C., and T. T. Kozlowski. 1973. *Ectomycorrhizae: Their Ecology and Physiology.* Academic, New York.

Marx, D. H., and S. V. Krupa. 1978. Ectomycorrhizae. In Y. Dommergues and S.

V. Krupa (Eds.), *Interactions between Nonpathogenic Soil Microorganisms and Plants*. Elsevier, Amsterdam, pp. 373–400.

Morgan, P. W., H. E. Joham, and J. V. Amin. 1966. Effect of manganese toxicity on the indole-acetic acid oxidase system of cotton. *Plant Physiol.* 41:718–724.

Morgan, P. W., D. M. Taylor, and H. E. Joham. 1976. Manipulations of IAA-oxidase activity and auxin deficiency symptoms in intact cotton plants with manganese nutrition. *Physiol. Plant.* 37:149–156.

Price, C. A. 1970. *Molecular Approaches to Plant Physiology*. McGraw-Hill, New York.

Robinson, A. D., and W. A. Hodgson, 1961. The effect of some amino acids on maganese toxicity in potato. *Can. J. Plant Sci.* 41:436–437.

Romheld, V., H. Marschner, and D. Kramer. 1982. Responses to Fe deficiency in roots of Fe-deficient plant species. *J. Plant Nutr.* 5:489–498.

Sanders, F. E., B. Mosse, and P. B. Tinker. 1976. *Endomycorrhizas*. Academic, New York.

Schutte, K. H., and M. Mathews. 1968. An anatomical study of copper-deficient wheat. *Trans. R. Soc. Africa* 38:183–200.

Smith, P. F. 1962. Mineral analysis of plant tissues. *Annu. Rev. Plant Physiol.* 13:81–108.

Ulrich, A. 1952. The physiological bases for assessing the nutritional requirements of plants. *Annu. Rev. Plant Physiol.* 3:207–228.

Wright, M. J. (Ed.), 1976. Plant adaptation to mineral stress in problem soils. Cornell Univ. Agr. Exp. Sta., Ithaca, NY.

6

SALT STRESS

Salinity has been an important factor in the history of mankind and in the agricultural systems upon which mankind has relied. Civilizations have been destroyed by the encroachment of salinity on the soils as a result of poor water management. If there is limited rainfall, salt is not leached out of the soil volume in which crop plants take root and yields are reduced as the salt concentration increases. Irrigation water contains dissolved salts that are concentrated as the water evaporates and build up in the soil over time. Vast areas of the land surface are rendered unfit for agriculture because of the present low tolerance of crop plants to salt and the lack of knowledge about the tolerance mechanisms available for plant breeders in the selection processes. In some areas with agricultural potential, two approaches may be taken. One approach is to try to find inexpensive and cost-efficient ways of desalting seawater to be used for irrigation purposes. Another approach is to select crop plants that can grow and survive under highly saline conditions.

Many of the areas of restricted use because of salt are located where there is abundant solar energy that can be used by the plants to produce high yields. It is necessary, therefore, to better understand the ways that plants can tolerate salinity so that more use can be made of lands that would otherwise be wasted under conditions of an increasing need for food and fiber by the increasing populations.

Plants are stressed in two ways in a high salt environment. In addition to the water stress imposed by the increase in osmotic potential of the rooting medium as a result of high solute content, there is the toxic effect of high concentrations of ions. Some plants have evolved mechanisms for dealing with these stresses and others can become adapted to them. But, the majority of crop plants are susceptible and will not survive under conditions of high salinity or will survive but with decreased yields.

In the saline environment there is a preponderance of nonessential over essential ions. The plant must absorb the essential ions from a diluted source in the presence of highly concentrated nonessential ions. By studying plants that usually grow in saline environments, it is possible to gain some knowledge of the mechanisms used to tolerate the conditions.

Native plants are frequently designated as halophytes or glycophytes by ecologists. Halophytes grow in saline soils, in high concentrations of salts, and are either facultative or obligate halophytes. Sometimes a finer classification is used by calling those that tolerate extreme salinity euhalophytes and those that tolerate only moderate salinity oligohalophytes. Glycophytes, or nonhalophytes, cannot grow in the presence of high concentrations of salts or possess some mechanism by which the protoplasm is not exposed to high salt concentrations.

Levitt (1980) chose to separate salt stress from ion stress. If the salt concentration is high enough to lower the water potential by 0.05 to 0.1 MPa then the plant is under salt stress. If the salt concentration is not this high, the stress is ion stress and may be caused by one particular species of ion. In a practical sense the salt concentrations in salt stress situations are much higher than required to lower the water potential by 0.1 MPa and the ion concentrations that cause ion stress are much lower. Some plants tolerate sodium salt stress (natrophilic plants) while others are not tolerant (natrophobic plants). Stress caused by calcium salts differentiates plants into calcicole (those tolerant of the stress) or calcifuge plants (those that are not tolerant). The relationship of tolerance to the salt in seawater becomes important under conditions that may not be relevant to crop plants unless one is considering domestication of plants that ordinarily grow in brackish water or seawater, or plants resulting from breeding programs for tolerance to seawater.

MECHANISMS OF TOLERANCE

What are the physiological characteristics related to the survival and productivity of plants in environments of high salinity? And what are the differences between salt-tolerant and salt-intolerant species? Some of the characteristics of halophytes that may give clues as to the physiological mechanisms involved in tolerance are:

1. Succulence that may lead to dilution of intracellular salt such as occurs in *Salicornia* sp.
2. The presence of salt excreting glands that reduce the concentration of salt in the plant, such as occur, for example, in species of *Frankenia*, *Cressa*, *Spartina*, and *Glaux*.
3. Development of small leaves, water storage hairs, and aerenchyma.
4. A water potential range from -2 to -4 MPa, which is equivalent to an ion concentration of 400 to 700 mM. (In the species *Salicornia rubra* the sodium and chloride ions make up 75 to 93% of the total osmotic potential.)
5. Species that rely mainly on a higher ion uptake and internal concentration of ions to maintain turgor.
6. Sometimes there is synthesis and accumulation of organic solutes that aids in the maintenance of turgor.

An understanding of the distribution of salt ions in the plant is necessary to understand tolerance. In the halophytes more than 90% of the sodium is in the shoot, with the highest concentration in the leaves. Translocation out of leaves is nil. When sodium is applied to leaves, it remains there and is not translocated to other parts of the plant. This behavior differs from that of potassium, which is readily mobile. Sodium also accumulates in the vacuoles of cells where concentrations may reach several thousand ppm. Scanning electron microscopy and experiments using precipitation of silver chloride point to the tonoplast as the line of demarcation between high and low sodium concentrations. Isolation of sodium in the vacuoles may help account for the fact that enzymes from halophytes are no more sensitive to sodium than those isolated from glycophytes, or, in some cases, more sensitive (Greenway and Osmond, 1972). At 300 mM salt and above the activity of many enzymes is inhibited by one-half or more. Other enzymes such as phosphohydrolases are insensitive to salt concentrations (Ting and Osmond, 1973).

The effect of salt stress on phosphorus metabolism varies with plant species and external phosphorus concentration in the rooting medium. Many data point to the fact that salinity damages mechanisms controlling intracellular phosphorus concentrations. Adverse effects on growth would be expected as a result and death may ensue. For example, in kidney bean there is a decrease in phosphorus, adenosine triphosphate, and energy available for the young leaves (Maas and Nieman, 1978). In addition, salt stressed plants often look like phosphorus–deficient plants with small, dark green leaves, decreased shoot-to-root ratios, decreased tillering, prolonged dormancy of lateral buds, delayed and reduced flowering, and fewer and smaller fruits (Hewitt, 1963).

Leaves of salt stressed plants frequently contain unusually high concentrations of sugars as a result of the effects of the stress on phloem translocation or on reduced sink size because of reduced growth (Gauch and Eaton, 1942; Nieman and Clark, 1976; Strogonov, 1962). What are the effects of salinity stress on phloem translocation? There must be a higher concentration of ions or other solute in the phloem sieve tubes than exterior to them so that they may maintain turgor under conditions of salt stress, but what is the effect on the water potential gradients that cause movement along the seive tubes? We can only speculate that there must be a significant effect. Under conditions of high transpiration rates, the water potential of leaves that supply the carbohydrate to the seive tubes might be reduced to a value lower than the water potential of phloem cells. In the latter case translocation would be reduced or cease completely.

Because of the nature of the selection processes in absorption across membranes in root cells and leaf cells, the entrance of sodium becomes limited (Greenway, 1973). As a result of the site of absorption, there is sometimes a high concentration of salt in the roots and sometimes in the stems of plants, but low concentrations in the leaves (Jacoby, 1965). Plants in general may be divided into salt excluders and salt nonexcluders because of differences in the selectivity of the membranes from plant species to plant species.

Under salt-stress conditions the osmotic potential of the soil solution is similar to that brought about by drought. Some of the symptoms of salt stress are those characteristic of water stressed plants. Although salt stressed plants are stunted, they are not wilted, which means that the cells must have water potentials that enable them to compete for water from the xylem. One of the ways the water potential may be lowered is by an increase in solutes.

An accumulation of organic solutes in response to salt stress may in reality be a response to water stress caused by the differential in water potentials between the plant and the soil solution. Carbohydrates, amino acids, or organic acids may accumulate. Oxalate may be as high as 50 to 100 meq/g dry weight, while in crassulacean acid metabolism (CAM) plants malate is the acid that accumulates. Since halophytes are often C4 plants, a shift from organic acid synthesis to amino acid synthesis may take place under salt stress.

In general, the contribution of carbohydrates to the osmoticum is not great enough to account for any significant adjustment in turgor under stressful conditions. What may be happening in the development of tolerance to salt stress is that reduction in growth and yield is caused by an osmotic regulation based on electrolytes absorbed against a concentration gradient causing a reduction in dependence on photosynthate and the synthesis of organic solutes to maintain turgor. The nontolerance or sensitivity of plants may be related to their inability to increase ion uptake under salt stress conditions or their inability to rapidly translocate ions to leaves and to compartmentalize them in leaf cells. If so, which parts of the sensitivity can be overcome by selection? Is ion compartmentation an inheritable trait?

BREEDING FOR SALT TOLERANCE

Epstein (1976) has stated that there is no fundamental biological incompatibility between plant life and saline conditions. Plants grow well in salt water where life originiated. The problem is with the crop plants, which have been selected for attributes other than those for salt tolerance. Almost without exception, cultivated varieties of crop plants are intolerant of salt concentrations equaling those found in seawater. Crops being tested for genetic selection potential are mostly cereals, such as rice, sorghum, and barley. Wheat and corn are not as promising. Salt pressure on tissues of tobacco, pepper, and alfalfa lines in tissue culture have resulted in selection of individual cells that are resistant and have been cultured to mature plants and propagated. Selection and domestication of some wild types are also being tried. Somers (1978) has listed several plants with potential as crops, among them *Atriplex patula* variety hastata, *Chenopodium album*, and some of the grasses that may be useful as forages.

To test breeding lines for salt tolerance, the test conditions must be defined. For convenience, Nieman and Shannon (1976) define salinity as the presence of excessive concentrations of soluble salts in the soil solution. The unit of measure of salinity is electrical conductivity expressed in deci-siemens/m at 25°C. The measurement is performed on a soil saturation extract. If one knows the moisture percentage of the soil, one can then calculate from the electrical conductivity of the bulk soil what the salinity of the soil is. Further calculations using the conversion factor of electrical conductivity times -0.036 MPa/dS/m gives the osmotic potential of the soil solution. The osmotic potential of the soil solution is a useful value with regard to the effect of salts on water absorption but disregards the effects of the specific ions in the solution. Techniques for evaluating large, segregating populations for salt tolerance are still lacking, however.

Plant breeders need to be aware of the factors that affect the tests for salinity tolerance. One of them is the plant growth stage. Tolerance changes throughout the life cycle. For example, sugar beet is tolerant most of the time but is susceptible during seed germination. Intolerance, however, may be more apparent than real and may be the result of the higher concentration of salts in the upper 2 to 3 cm of soil because of upward movement of the soil solution and evaporation of water from the surface of the soil leaving behind a salt deposit. Rice, tomato, wheat, and barley exhibit reduced tolerance during the seedling stage of development.

Various factors in the atmospheric environment also affect the test. High atmospheric humidity increases tolerance, probably by reducing transpiration rates (Hoffman and Rawlins, 1971; Hoffman et al., 1971; Nieman and Poulsen, 1967). Plants are more tolerant at moderate temperatures than at high temperatures (Francois and Goodin, 1972; Magistad et al., 1943). Tolerance is also reduced by high light intensity (Nieman and Poulsen, 1971), and low oxygen concentrations are synergistic with salinity in reducing growth under saline conditions (Aceves et al., 1975). On the other hand, salt stress increases resistance to ozone damage, possibly by causing the stomates to close (Hoffman and Rawlins, 1971).

SPECIFIC ION EFFECTS

Some of the specific ion effects are discussed in Chapter 5. Some aspects of sodium chloride stress need to be dealt with at this time. The rate of uptake of ions by roots differs as a result of differences in membrane permeability, transport kinetics and selectivity, and feedback controls. Chloride ions contribute more to an osmotic adjustment than do sulfate ions because chloride is more readily absorbed. In the presence of predominantly monovalent cations and divalent anions, the cation uptake exceeds that of the anions (Maas and Nieman, 1978). A balance of charges is achieved by the synthesis of organic anions within the cell or tissue. If the inorganic divalent anions are nitrate and sulfate, the balance is further disturbed by the metabolism of nitrate and sulfate.

Sodium ions in a saline soil may exceed those of potassium by more than two orders of magnitude but the sodium to potassium ratio in plants is nearly 1:1. The root cells have a highly specific mechanism for absorbing potassium to the exclusion of sodium or else the cells secrete the sodium and retain more of the potassium. Some woody plants are susceptible to chlorine and their tolerance of saline conditions depends on their tolerance to chlorine and how much of the ion is absorbed and translocated.

DISCUSSION QUESTIONS

1. What are the most prominent causes of saline conditions in agricultural environments?

2. Why should you, as a student, be concerned about crop plant tolerance of saline conditions?

3. Name some of the factors causing salt stress of plants and explain how each might be a cause of reduction in growth and yield.

4. What is a halophyte?

5. What are the physiological responses of plants to salt stress? Separate them into those that happen shortly after the onset of stress and those that develop after several weeks have elapsed.

6. If a tolerant species or cultivar is growing in a saline environment adjacent to a glycophyte, what are the differences one would expect

to find in the physiology of the two plants? Which features would be the same?

7. How are plants able to maintain turgor? How do the phloem seive tubes maintain turgor when the plant is subjected to salt stress?

8. How would you proceed to distinguish between symptoms of phosphorus deficiency and salt stress?

9. Describe the effects of salinity stress on translocation.

10. What are the most productive approaches to breeding for tolerance to salt stress? How do you know?

REFERENCES

Aceves, N. E., L. H. Stolzy, and G. R. Mehuys. 1975. Combined effects of low oxygen and salinity on germination of a semi-dwarf Mexican wheat. *Agron. J.* 67:530–532.

Aller, J. C., and O. R. Zaborsky. 1979. The biosaline concept. In A. Hollaender et al. (Eds.), *The Biosaline Concept*. Plenum, New York.

Bernstein, L. 1975. Effects of salinity and sodicity on plant growth. *Annu. Rev. Phytopathol.* 13:295–312.

Boyko, H. (Ed.). 1966. *Salinity and Aridity: New Approaches to Old Problems*. Junk, The Hague.

Boyko, H. (Ed.). 1968. *Saline Irrigation for Agriculture and Forestry*. Junk, The Hague.

Epstein, E. 1976. Genetic potentials for solving problems of soil mineral stress: Adaptation of crops to salinity. In M. J. Wright (Ed.), *Plant Adaptation to Mineral Stress in Problem Soils*. Cornell Univ. Press, Ithaca, NY. pp. 73–82.

Flowers, T. J., P. F. Troke, and A. R. Yeo. 1977. The mechanism of salt tolerance in halophytes. *Annu. Rev. Plant Physiol.* 28:89–121.

Francois, L. E., and J. R. Goodin. 1972. Interaction of temperature and salinity on sugar beet germination. *Agron. J.* 64:272–273.

Gauch, H. G., and F. M. Eaton. 1942. Effect of saline substrate on hourly levels of carbohydrates and inorganic constituents of barley plants. *Plant Physiol.* 17:347–365.

Greenway, H. 1973. Salinity, plant growth and metabolism. *J. Austr. Inst. Agr. Sci.* 39:24–34.

Greenway, H., and R. Munns. 1980. Mechanisms of salt tolerance in nonhalophytes. *Annu. Rev. Plant Physiol.* 31:149-190.

Greenway, H., and C. B. Osmond. 1972. Salt responses of enzymes from species differing in salt tolerance. *Plant Physiol.* 49:256-259.

Hellebust, J. A. 1976. Osmoregulation. *Annu. Rev. Plant Physiol.* 27:485-505.

Hewitt, E. J. 1963. The essential nutrient elements: Requirements and interactions in plants. In F. C. Steward (Ed.), *Plant Physiology, a Treatise*. Vol. 3. Academic, New York, pp. 137-360.

Hoffman, G. J., and S. L. Rawlins. 1971. Growth and water potential of root crops as influenced by salinity and relative humidity. *Agron. J.* 63:877-880.

Hoffman, G. J., S. L. Rawlins, M. J. Garber, and E. M. Cullenm. 1971. Water relations and growth of cotton as influenced by salinity and relative humidity. *Agron. J.* 63:822-826.

Hoffman, G. J., E. N. Maas, and S. L. Rawlins. 1975. Salinity ozone interactive effects on alfalfa yield and water relations. *J. Environ. Qual.* 41:326-331.

Hollaender, A., J. C. Aller, E. Epstein, A. san Pietro, and O. R. Zaborsky (Eds.), 1979. *The Biosaline Concept: An Approach to Utilization of under Exploited Resources*. Plenum, New York.

Jacoby, B. 1965. Sodium retention in excised bean stem. *Physiol. Plant.* 18:730-739.

Levitt, J. 1980. *Responses of Plants to Environmental Stresses*. Vol. 2. Academic, New York, pp. 365-488.

Maas, E. V., and R. H. Nieman. 1978. Physiology of plant tolerance to salinity. In G. A. Jung et al. (Eds.), *Crop Tolerance to Suboptimal Land Conditions*. Am. Soc. Agron., Madison, WI, pp. 277-299.

Magistad, O. C., A. D. Ayers, C. H. Wadleigh, and H. G. Gauch. 1943. Effect of salt concentration, kind of salt, and climate on plant growth in sand culture. *Plant Physiol.* 18:151-166.

Mudie, P. J. 1974. The potential economic uses of halophytes. In R. J. Reimold and W. H. Queen (Eds.), *Ecology of Halophytes*. Academic, New York, pp. 565-597.

Nieman, R. H., and L. L. Poulsen. 1967. Interactive effects of salinity and atmospheric humidity on the growth of bean and cotton plants. *Bot. Gaz.* 128:69-73.

Nieman, R. H., and R. A. Clark. 1976. Interactive effects of salinity and phosphorus nutrition on the concentrations of phosphate and phosphate esters in mature photosynthesizing corn leaves. *Plant Physiol.* 57:157-161.

Nieman, R. H., and L. L. Poulsen. 1971. Plant growth suppression on saline media: Interactions with light. *Bot. Gaz.* 132:14-19.

Nieman, R. H., and M. C. Shannon. 1976. Screening plants for salinity tolerance. In M. J. Wright (Ed.), *Plant Adaptation to Mineral Stress in Problem Soils*. Cornell Univ. Press, Ithaca, NY, pp. 359–372.

Pasternak, D., M. Tervrsky, and Y. deMalach. 1979. Salt resistance in agricultural crops. In H. Mussell and R. C. Staples (Eds.), *Stress Physiology in Crop Plants*. Wiley-Interscience, New York, pp. 128–142.

Polkajoff-Mayber, A., and J. Gale (Eds.). 1975. *Plants in Saline Environments*. Springer-Verlag, Berlin.

Rains, D. W. 1972. Salt transport by plants in relation to salinity. *Annu. Rev. Plant Physiol.* 23:357–388.

Reimold, R. J., and W. H. Queen. 1974. *Ecology of Halophytes*. Academic, New York.

Rush, D. W., and E. Epstein. 1976. Genotypic response to salinity: differences between salt sensitive and salt tolerant genotypes of tomato. *Plant Physiol.* 57:162–166.

Somers, G. F. 1978. Production of food plants in areas supplied with highly saline water: Problems and prospects. In H. Mussell and R. C. Staples (Eds.), *Stress Physiology in Crop Plants*. Wiley-Interscience, New York, pp. 108–125.

Strogonov, B. P. 1962. Physiological basis of salt tolerance of plants. (Transl. from Russian by A. Poljakoff-Mayber and A. M. Mayer) Israel Program Sci. Transl., Jerusalem, 1964.

Ting, I. P., and C. B. Osmond. 1973. Photosynthetic phosphoenolpyruvate carboxylases. Characteristics of alleloenzymes for leaves of C3 and C4 plants. *Plant Physiol.* 51:439–447.

7

IRRADIATION STRESS

All life on earth is supported by the radiant energy from the sun, which plants convert into chemical energy by the process of photosynthesis. Light is one of the most important and variable components of the plant environment. The autotrophic plant is directly influenced by the intensity of light, which drives photosynthesis, and thereby, provides the chemical energy and carbon needed for plant growth and development. Thus, an alteration in light intensity, whether a deficit or excess, will result in a disruption of plant metabolic processes. Besides affecting photosynthetic rate, solar radiation also affects plant temperature and photomorphogenic responses in ways that can produce stress.

ATMOSPHERIC ATTENUATION OF SOLAR RADIATION

Light quality and quantity perceived by the plant are attenuated in a number of different ways as the solar radiation enters the earth's atmosphere. Unfiltered sunlight entering the ionosphere has an intensity of 1.39 kW/m^2 (the solar constant) and wavelengths from 225 to 3200 nm, with 41% of this radiation between the wavelengths of 400 and 700 nm, the physiologically active wavelengths in plants. The ozone layer in the strato-

TABLE 7.1 Characteristics of Various Wavelength Regions of Light

Color	Wavelength Range (nm)	Representative Wavelength (nm)	Frequency (cycles/s) (Hz × 10^14)	Energy (eV/photon)	kcal/mol of Photons
Ultra-violet	Below 400	254	11.80	4.88	112.5
Violet	400–425	410	7.31	3.02	69.7
Blue	425–490	460	6.52	2.70	62.2
Green	490–560	520	5.77	2.39	55.0
Yellow	560–585	580	5.17	2.14	49.3
Orange	585–640	620	4.84	2.00	46.2
Red	640–740	680	4.41	1.82	42.1
Infrared	Above 740	1400	2.14	0.88	20.4

sphere absorbs ultraviolet radiation while water vapor, carbon dioxide, and oxygen in the trophosphere absorb the wavelengths of 1100 to 3200 nm. At sea level range of light wavelengths available to the plant is from 310 to 1100 nm with 46% of the radiation between wavelengths of 400 to 700 nm (visible light). Thus, of the solar radiation available from the solar constant, only 47% reaches the earth's surface. Over half is lost by refraction and diffraction in the high atmosphere. Clouds and particulates suspended in the air also reflect, scatter, or absorb the sun's radiation. At noon a total radiation of up to 900 W/m² can be received by a plant at sea level at an intermediate latitude. The amount of radiation varies depending upon cloud cover, latitude, and altitude. Thus, the solar radiation available to the plant is greatly attenuated by various factors in the environment. For this reason light stress is usually due to a deficiency rather than to an excess of light. The characteristics of various wavelengths of light are given in Table 7.1.

DISTRIBUTION OF RADIATION IN A PLANT COMMUNITY

Solar radiation is attenuated not only by the atmospheric conditions but also by the plant habitat. In the plant community, photosynthesizing leaves form an overlapped and stacked arrangement. The incident light in these stands is progressively absorbed by the stacked leaves so that most of the light impinging upon a plant is utilized. The quantity and quality of light available to plant leaves will depend on the stand density, plant

TABLE 7.2 Summary of Photoreactions of Plants

Photo-process	Reaction or Response	Photo-receptor	Action Spectra Peaks (nm)	Energy Conversion or Products
Chlorophyll synthesis	Reduction of proto-chlorophyll	Proto-chlorophyll	Blue: 445 Red: 650	Chlorophyll *a* Chlorophyll *b*
Photo-synthesis	Dissociation of water	Chlorophylls Carotenoids	Blue: 435 Red: 675	Reductant (H) Phosphorylated compounds
	Carbon reduction enhancement	Chlorophylls	Red: 650 Far red: 710	Phosphorylated compounds
Blue light reactions	Phototropism	Carotenoids Other flavins	UV: 370 Blue: 445,475	Oxidized auxin, auxin system, other components of the cells
	Protoplasmic viscosity	Unknown	Uncertain	
	Photo-reactivation	Pyridine nucleotide, riboflavin	Uncertain	
Red, far-red reactions	Seed germination Growth	Phytochrome	Induction by red: 660	Research continues
	Anthocyanin synthesis	Phytochrome	Reversed by far-red:	
	Chloroplast responses	Phytochrome	710 & 730	
Other: heterotrophic growth, photoperiodism				

Source. R. B. Withrow. 1959. A kinetic analysis of photoperiodism. In R. B. Withrow (Ed.), *Photoperiodism and Related Phenomena in Plants and Animals*. Used with permission from the American Association for the Advancement of Science, Washington, DC.

and leaf shape. More light will be available to leaves at lesser
in stands that contain plants with narrow leaves such as grain
meadows, or reed clumps compared to communities containing
with broad leaves. Forests with closely packed tree crowns and
oliage will intercept much more radiation than open stands of trees
species with sparse crowns.

t of several distinct wavelength ranges affects different photopro-
nd is absorbed by recognizable photoreceptors. A summary of the
actions of plants and the action spectra is given in Table 7.2.

UPTAKE OF RADIATION BY PLANTS

Plants can either reflect, absorb, or transmit the solar radiation incident on their leaves. The degree of these three reactions will depend on the wavelength of the radiation, leaf structure, and leaf orientation. The capacity of a plant to reflect visible light will depend on the leaf surface. Modifications, such as hairs, will increase reflection of solar radiation. Leaves can reflect 70% of the infrared radiation, 6 to 12% percent of the visible radiation, but only 3% of the ultraviolet radiation. Green light tends to be reflected more strongly (10 to 20%) than orange or red light (3 to 10%).

Radiation that penetrates the leaf can be absorbed by various components. Ultraviolet light is greatly absorbed by epidermal wax, cuticle, and suberin as well as by phenolic compounds within the leaf. Chloroplast pigments determine the extent of visible light absorption. Infrared radiation above 700 nm (heat radiation) is readily absorbed by the plant. Transmission of radiation depends mainly on leaf structure and thickness, thus, thin leaves will transmit more radiation than thick leaves. Wavelengths that are reflected tend to be more extensively transmitted with the result that there is an enrichment of the wavelengths around 500 nm and above 800 nm.

SUN VERSUS SHADE PLANTS

In low light, photosynthesis is linearly dependent on light intensity where utilization (quantum yield) is at a maximum. At higher light intensities, photosynthesis becomes less than proportional and eventually does not increase with increased light intensity. The reaction of a plant to different intensities of solar radiation will differ depending on whether the plant is a shade or sun plant. Plants occupying sunny habitats (sun plants) require high light intensities to maintain a high photosynthetic rate and, consequently, show lower rates of photosynthesis at low light intensities. Plants occupying shaded habitats (shade plants) are capable of greater photosynthetic rates in low light intensities. Sun plants also have a higher light compensation point than shade plants. For the most part, the success of a plant depends on its ability to continue to photosynthesize under low light conditions. Typically, at light intensities below the compensation point, injury will occur in the plant as a result of starvation. The most rapid effect is a decrease in carbohydrate content followed by other alterations of metabolism. Because of the low rate of dark respiration in shade plants, low light intensities are sufficient to reach the light compensation point. Sun plants have higher respiration rates and require greater light intensities to reach the compensation point. Light saturation is reached early in shade plants at approximately 5% of maximum daylight, while sun plants continue to respond linearly to higher light intensities. Adaptive characteristics permit shade plants to function under low light intensities occurring in their environment and enable sun plants to effectively use moderate and high light intensities. A summary of the range of intensities found in the natural environment and the physiological processes that are affected by the different intensities is given in Figure 7.1.

A major difference between sun plants and shade plants occurs in the chloroplasts. Shade plants have large grana stacks with approximately 100 thylakoids per granum oriented irregularly with the chloroplast. There is a greater proportion of lamellae-forming grana and a larger ratio of thylakoid membranes to stroma with the result that there is a greater chlorophyll content per unit leaf area and a lower ratio of chloroplasts per unit leaf area in shade plants than in sun plants . Chloroplasts in sun plants, on the other hand, have grana oriented in one plane and a lower proportion of lamellae-forming grana and a lower ratio of thylakoid membranes to stroma. However, sun plants grown in low light have increased thylakoid stacking and thylakoid to stroma ratios.

Sun and shade plants also vary in the concentration of chlorophyll present and in the ratio of chlorophyll *a* to chlorophyll *b* in the chloroplasts. The light absorption efficiency of a plant depends on the amount of chlorophyll per unit leaf area that, for a particular plant, remains constant over wide ranges of light intensities. The chlorophyll concentration and ratio of cholorophyll *a* to chlorophyll *b* of a plant can, however, adapt to changing environmental conditions. Severe shading may cause a decrease in chlorophyll concentration that, in sun plants, is more pronounced in young developing leaves. On the other hand, obligate shade plants in deep shade can have as much chlorophyll as sun plants grown at high light intensities.

Under low light intensities the chlorophyll *b* concentration increases in most plants whether they be shade or sun plants. The shift allows the capture and use of filtered light containing wavelengths that are enriched in

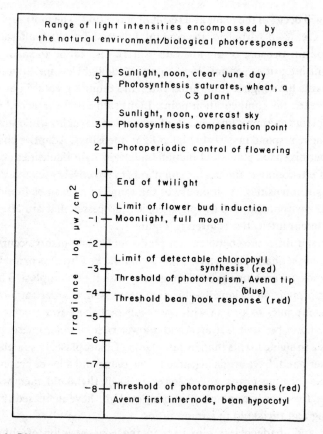

Figure 7.1. Diagram of the light intensities at which various plant reactions occur. From R. B. Withrow, 1959. A kinetic analysis of photoperiodism. In R. B. Withrow (Ed.), *Photoperiodism and Related Phenomena in Plants and Animals*. Used with permission of the American Association for Advancement of Science, Washington, DC.

the far-red region of the spectrum. In addition, shade plants normally have lower chlorophyll *a* to chlorophyll *b* ratios than sun plants. Since chlorophyll *b* is a part of the light harvesting complex (LHC), differences in the ratio may be reflected in the proportion of LHC to total chlorophyll content. The light harvesting complex is associated with photosystem II (PS

II), thus, shade plants may have a greater PS II to PS I ratio, enabling them to compensate for increases in the far-red light filtered through the canopy.

In contrast to sun plants, shade plants contain less protein per unit of chlorophyll and per unit of leaf area. The reduced protein level is brought about mainly by the reduced amount of ribulose bisphosphate carboxylase and other enzymes of photosynthetic carbon metabolism. Shade plants also have lower nonsoluble protein content associated with the chloroplast membrane.

The accumulation and distribution of dry matter in sun and shade plants also varies. Productivity under low light intensity depends on net photosynthesis, the leaf area intercepting and trapping solar energy, and the allocation of photosynthate to photosynthetic and nonphotosynthetic tissues.

EFFECTS OF LIGHT DEFICIT (SHADE)

Since plants cannot move about they must adapt to new light intensities as they occur. Light intensities below the compensation point lead to starvation with carbohydrates used as a substrate for respiration first and then other sources of energy. Adaptation to shade comes about in two ways: (1) increase of leaf area in a way that minimizes the use of metabolites (for example, by increasing leaf area at the expense of photosynthate allocation to root growth), and (2) a decrease in the amount of transmitted and reflected light. Shade leaves are thinner but larger in surface area than sun leaves. Increased light absorption is brought about by increased number of chloroplasts per unit leaf area and by increased chlorophyll concentration of the chloroplasts which is accompanied by a lowered concentration of other pigments that interfere with the light absorption process. In addition, there is sometimes a change in the orientation of the chloroplasts so that the broad dimension is oriented toward the light.

Resistance to chlorophyll degradation in shade plants serves to increase chlorophyll concentration. In some plants there is also a reduction in the respiratory rate, which conserves photosynthate.

EFFECTS OF BRIGHT LIGHT

The photosynthetic performance of plants in high intensity light is not dependent on the chlorophyll content or on absorption and trapping of light quanta as in shade plants but rather on the relationship between photoinhibition of photosynthesis and the degree of chlorophyll destruction. The extent of photoinhibition is dependent on the intensity and duration of exposure to high light intensities. Photoinhibition is caused by a transfer of the excess quanta produced during the light reaction to the reaction centers instead of being disposed of as heat or fluorescence. Creation of singlet oxygen or free radicals destroys the reaction centers unless a receptor for the increased reducing power is present (see Chapter 2). Usually the receptor is the carbon dioxide fixation processes involving RuBPcase (ribulose bisphosphate carboxylase) that use the quanta. Shade plants have low levels of RuBPcase while sun plants, especially C3 plants, have higher levels of the enzyme (Chapter 2). Sun plants with C4 metabolism are able to use photosynthetic energy more efficiently without an increase in RuBPcase.

Obligate shade plants have a lower capacity for electron transport brought about by lower content of the carriers cytochrome f, cytochrome $b559$ and $b563$, plastoquinone, ferredoxin, and ATP synthetase. Injury to such plants by high light intensities is dependent upon the concentration of carbon dioxide and oxygen present as receptors of the increased reducing power generated. Low temperatures cause a reduction of carbon dioxide fixation again decreasing the availability of receptors of reducing power.

Leaf modifications that influence photosyhthesis will also alter the plant response to radiation stress. Stomatal conductance, for example, will affect the rate of absorption of carbon dioxide. Leaf growth and development is changed by high light intensities in such a way that there is an increase in the elongation of the palisade cells and an increase in the number of cells across the leaf section and in the average cell diameter.

RESISTANCE TO HIGH LIGHT INTENSITY INJURY

Shade plants have a limited capacity for adjustment to high light intensities and are usually injured or killed. Sun plants, on the other hand, gradually increase the capacity for light saturated photosynthesis by (1) increasing the level of carbon metabolism enzymes, (2) increasing photosynthetic electron transport chain carriers, or (3) increasing the capacity for carbon dioxide transport. Resistance to high light stress also occurs in a number of other ways. Increased suberin, wax, cuticle, and cell wall materials of the leaves can cause an increase in the amount of reflected light. The photodestruction of chlorophyll can be reduced by an increase in the content of β-carotene, xanthophyll, and phospholipid of the leaves. An increase in the stacking of grana accompanied by a stabilization of the chloroplast membranes aids in preserving chlorophyll. And, the heliotropism of the leaves can cut down on the amount of light intercepted.

ULTRAVIOLET RADIATION

The electromagnetic radiation from the sun contains about 7% ultraviolet at sea level. Ultraviolet radiation is abosrbed by plant chromophores, which include nucleic acids, proteins, indolacetic acid, abscisic acid, and flavoproteins. The role of chromophores has not been well established and their identification has been complicated by the secondary effects occurring in plants in response to irradiation by ultraviolet. Absorption by nucleic acids may alter protein synthesis but proteins are also absorbers of radiation and may react at a primary level. Alteration of protein and lipid membrane components by ultraviolet may alter membrane permeability and ionic balance. Such events also cause an inhibition of photosynthesis and respiration.

The most common inactivation of nucleic acids by ultraviolet radiation is through photochemical lesions involving polymers of pyrimidine bases in the deoxyribonucleic acids (DNA). The result is the production of pyrimidine dimers and loss of DNA biological activity. It is evident that ultraviolet absorption can cause an acceleration in the mutation rates and aberrations of chromosomes. Repair systems for the DNA molecules have been found in plants, however, and involve the enzymatic splitting of the dimers formed by ultraviolet absorption. Such a repair system has been implicated in repairing epidermal tissue damage, restoration of growth rate, and synthesis of chlorophyll and of nitrate reductase.

In some cases, repair of ultraviolet radiation damage can be accomplished without light activation. The repair system consists of enzymatic reactions in which the DNA lesions are excised from the DNA strand and replaced by newly synthesized pieces. Or, undamaged DNA is replicated with gaps in the place of the lesions and a patch is synthesized from information within the cell and spliced into the new DNA strand. Unlike DNA, ribonucleic acid (RNA) and protein are better able to withstand ultraviolet radiation damage.

Reduced plant growth with exposure to ultraviolet radiation can occur as a result of inhibition of photosynthesis or a reduction in leaf expansion. Since both indole acetic acid and abscisic acid are receptors for ultraviolet radiation, alterations of concentrations of these growth regulators may result in irregular growth. Such irregularities may take the form of depressed flower development, loss of apical dominance, abscission of leaves, and alteration of mineral nutrient concentrations in tissues.

Resistance to ultraviolet radiation damage consists of the ability of a plant to repair the damage to the DNA, or to synthesize screening compounds, such as flavanoids and flavones, in the epidermis. Cuticular waxes may also be effective in absorbing harmful ultraviolet radiation. Changing the angle of the leaves to the incidence of radiation or increasing the reflecting properties of leaf surfaces are other mechanisms of resistance.

IONIZING RADIATION AS STRESS

Ionizing radiations are radiations that produce ions in the media they traverse. Energy is transferred from the moving particles and causes the ejection of an electron or of a proton or capture of the particle by the nucleus of the atom that gets in the way. Charged particles directly produce many ionizations per particle. Photons in X rays, γ rays, or neutrons produce few energetic charged particles but each in turn produce many ionizations. In biological materials the production of electrons causes all of the ionizations. Excitation of matter is considered relatively unimportant in causing injury in biological material. It is the free radicals formed that are injurious.

Free radicals contain unpaired electrons and are extremely reactive in chemical reactions within the system in which they are produced. The resultant injury to the plant may arise from alteration of essential components of protoplasm such as the genetic material in which mutations can be created. Alterations, deletions, and additions to the DNA acid bases can occur. Such changes in the genetic potential of cells has been used in some breeding programs to increase the rates of mutations.

The symptoms of radiation damage in plants take many forms. A change in ploidy with the formation of polyploids can lead to loss of reproductivity, mitotic delay, lengthened cell cycle time, and change in cell morphology. The processes underlying differentiaton may be changed, resulting in gall formation or even death. At the whole plant level, the symptoms take the form of early senescence, somatic mutations, increased photosensitivity of seed germination, change in resistance to stress, prevention of geotropic responses, production of ethylene, and many others depending upon the condition of the plant at the time of exposure and the amount or mass of dead tissue present as in the woody species. Some of the physiological factors that determine the extent of radiation damage are water content, ambient temperature, oxygen concentration, and the state of nutrition.

Seeds are more susceptible to radiation when they have a high water content. Hydrated seeds have a low decay rate of the radiation-induced radicals. When oxygen is present there is an increase of free radical production. Phosphorous and calcium deficiencies enhance radiosensitivity as does the presence of boron. Temperature has some effect on the ability of the plant to repair ionization radiation damage because low temperatures slow or prevent repair while relatively high temperatures enhance the recovery processes.

DISCUSSION QUESTIONS

1. Discuss how solar radiation entering the earth's atmosphere is attenuated before being received by the plant.

2. In what ways can a plant community alter solar radiation?

3. The uptake of radiation by plants can occur in three ways. What are they? How do they affect the plant? What modifications can be made to alter a plant's uptake of radiation?

4. Contrast and compare the morphology of sun and shade plants.

5. What is the light compensation point? How do sun and shade plants vary their responses to different light intensities and compensation points?

6. How do sun and shade plants vary in their metabolic requirements? How does this alter their ability to withstand deficit light stress?

7. What are the two ways in which a plant can adapt to shade? Give an example of each adaptation.

8. What is photoinhibition and how does it occur?

9. How do bright light intensities affect a plant's chlorophyll content? What environmental factors alter the plant response to bright light?

10. List the ways in which a plant can resist high light intensity injury.

11. What are some of the chromophores in a plant that can absorb ultraviolet radiation?

12. Give an outline of what would happen in a plant's metabolism if destruction of nucleic acids by ultraviolet radiation occurred as a first step. What differences would occur if the proteins were destroyed first?

13. Describe two ways in which DNA can be repaired after ultraviolet radiation damage. What are some of the other protective properties that plants develop?

14. What are some of the effects of ionizing radiation on a plant's genetic information? How can these genetic alterations be expressed at the cell, tissue, and organ level?

15. Sensitivity and modification to ionizing radiation in plants depends on what factors?

REFERENCES

Bjorkman, O. 1981. Responses to different quantum flux densities. In O. L. Lange, P. S. Nobel, C. B. Osmond, and H. Ziegler (Eds.), *Interactions with the Physical Environment*, Vol. 12A. Springer-Verlag, Berlin, pp. 57–107.

Boardman, N. K. 1977. Comparative photosynthesis of sun and shade plants. *Annu. Rev. Plant Physiol.* 28:355–377.

Critchley, C. 1981. The mechanism of photoinhibition in higher plants. In G. Akoyunglou (Ed.), *Photosynthesis.*, Vol. 6, Proceedings, Fifth Photosynthesis Congress, Balaban International Science Services, Philadelphia, pp. 297–305

Grime, P. J. 1966. Shade avoidance and shade tolerance in flowering plants: Effects of light on the germination of species of contrasted ecology. In R. Bainbridge, G. C. Evans, and O. Rackman (Eds.), *Light as an Ecological Factor*. Blackwell, Oxford, pp. 187-207.

Haber, A. H. 1968. Ionizing radiation as a research tool. *Annu. Rev. Plant Physiol*. 19:463-482.

Larcher, W. 1980. *Physiological Plant Ecology*. 2nd ed. Springer, New York.

Levitt, J. 1980. *Responses of Plants to Environmental Stresses*. Vol. 2, Academic, New York, pp. 283-364.

Smith, H. 1975. *Light and Plant Development*. Butterworths, Boston.

Smith, H. (Ed.). 1981. *Plants and the Daylight Spectrum*. Academic New York.

Withrow, R. B. 1959. A kinetic analysis of photoperiodism. In R. B. Withrow (Ed.), *Photoperiodism and Related Phenomena in Plants and Animals*, Am. Assoc. Adv. Sci., Washington, DC. pp. 439-471.

8

ALLELOCHEMICAL STRESS

The biological entities in the environment can cause stress by competing for nutrients, light, water, and space but they also often release chemical compounds that affect their progeny or other plants in ways that are dependent upon concentration of the chemicals released, the nature of the chemicals, and the period of time the chemicals are in the environment. Molisch (1937) is credited with coining the word "allelopathy" for the chemical interactions that occur among living organisms. The organic compounds involved in allelopathy are called allelochemicals.

Allelochemicals become stressful only when they are toxic or when they affect the growth and development of plants in such a way as to render them more susceptible to other environmental stresses. Allelochemicals may also be of benefit to plants and frequently the difference between beneficial and detrimental effects is one of concentration of the allelochemical in the environment and the length of time a plant may be exposed to it (Rice, 1974). Specific terms for the allelochemical interactions between one microorganism and another, microorganisms and plants, and one plant and another have been coined by various scientists and may be found in the book on allelopathy by Rice (1974).

Putnam and Duke (1978) proposed that allelopathic plants, or the allelochemicals from such plants, could be used in agricultural ecosystems to manage plant pests in a variety of ways:

1. By interplanting allelopathic plants with crop plants.
2. By increasing the allelopathic potential of crop plants through selection and breeding.
3. By using allelopathic plants as sources of pesticides.
4. By eliminating allelopathic plants (weeds) that inhibit crop plants.
5. By identifying and synthesizing new compounds that appear as secondary metabolic compounds in allelopathic plants.

However, the interactions may be quite complex and the interfering chemicals may not come directly from a single organism or plant but may arise as a result of decomposition processes and formation of humus in soil. Allelopathy is very apt to be the result of leaching and litter decompostion (Kaminsky, 1981). There are many reports in the literature of allelopathic effects of the addition to soil of ground leaf litter, extracts of whole plants, or washings from leaf surfaces. But, more and more there are reports of quantitative effects of individual allelochemicals and the identification of the allelochemicals involved in defined ecological situations.

THE JUGLONE STORY

Perhaps one of the best known examples of allelopathy is the effect of black walnut trees on herbaceous crops planted under the canopy of the tree or over the root system. In a study by Massey (1925) it was demonstrated that alfalfa plants were killed for a radial distance well exceeding the spread of the branches but not the extent of the roots. The area subsequently became inhabited by grasses. In a separate study tomato plants in pots were placed at varying distances from the trunk of the walnut tree and left for a period of time. The leaf drippings were toxic to the tomatoes. Similar studies of walnut trees near apple trees (Schneiderman, 1927) indicated that roots of walnut and apple trees came into intimate contact resulting in the death of the apple tree roots and of the above-ground side of the tree situated closest to the walnut tree. Many other observations of a

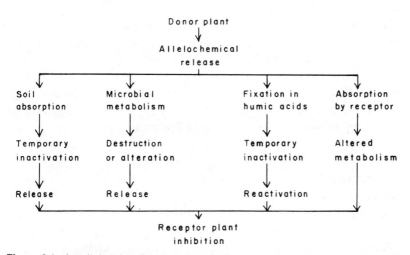

Figure 8.1. An alleleochemical may be altered in a variety of ways as it moves from the donor plant in which it originated to a receptor plant where it may act as an inhibitor of metabolism. From S. B. Horsley, 1976. Allelopathic interference among plants, II. Physiological modes of action. In H. E. Wilcox and A. T. Hamer (Eds.), *Proceedings of the 4th North American Forest Biology Workshop.* Used with permission of the U. S. Forest Service.

similar nature have been recorded. Release of a toxic substance from the roots of walnut trees and from the leaves in the drip zone was suspected. Davis (1928) reported the identification of the compound, 5-hydroxy-α-naphthaquinone, from extracts of leaves, roots, and hulls of walnut and named it juglone after the generic name of walnut, *Juglans.* Other species of juglans have also been shown to release juglone into the environment. Research is currently underway on the interaction of black walnut with other species in plantings from which the walnut is to be used for furniture wood and for the nuts.

The juglone story is an example of how observations not explainable by competition were later proven to be caused by a toxic substance released in any of a number of ways into the environment of other plants causing changes in the pattern of vegetation and loss of valuable crops. The example of a chemical stress caused by release of a toxin from a donor plant can be cited for many other situations. Some of the possible ways in which an allelochemical may be altered in its pathway to a donor organism are outlined in Figure 8.1. In this case the final receptor is another plant.

SOURCES AND NATURE OF ALLELOCHEMICALS

Since allelopathy implies that chemical compounds are released into the environment, how do they get there? The release of allelochemicals from plants occurs by a variety of means such as volatilization, leaching by fog drip or rain, exudation from roots, or degradation of dead plant parts. All plant parts have been shown to contain inhibitors. Leaves, stems, roots, rhizomes, flowers, fruits, and seeds have been bioassayed and found to contain inhibitors but leaves and roots are the most important sources. Many chemicals are localized in plant organs so that it is not unusual to find different allelopathic compounds in different parts of the same plant.

Large quantities of water-soluble organic constituents can be leached from living tissues by rainwater or fog-drip (del Moral and Muller, 1969; Tukey, 1969, 1970; Tukey and Mecklenburg, 1964). As tissues approach senescence they are more prone to leaching of metabolites than young tissues because of conditions that favor wettability (Tukey, 1969). Factors of the environment that affect plant development may also affect leaching by changing the wettability of surfaces and movement of organic molecules to the surfaces.

A wide variety of organic compounds are known to be released from roots into the rhizosphere. The roles of chemicals in interactions among organisms in the rhizosphere have been classified into the general categories of nutrients, foods, and allelochemicals (Whittaker and Feeney, 1971). The amounts released vary with a number of factors that include the usual loss of cells and tissues from the root surface and root cap as the root grows in length and diameter, injury from a variety of sources, stage of development of the root, its type and growth rate, and all the environmental factors that affect root processes and the activity of microorganisms in the rhizosphere (Hale and Moore, 1979). Release of allelochemicals follows the patterns of diffusion, secretion, or sloughage from the root surface into the surrounding soil. The difficulty in allelopathic studies is in knowing whether a particular inhibitor was actually exuded, whether it came from dead cells sloughing off the root, or whether it was a transformation product of rhizosphere microorganisms (Rovira, 1969).

Most evidence for root-mediated allelopathic effects has come from studies in which root washings have been produced by recycling distilled water or nutrient solution through interconnected pots of sand or soil containing the potential toxin-producing and test plants (Horsley, 1976). Such

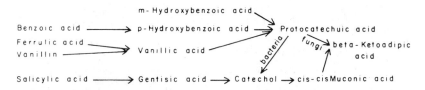

Figure 8.2. Growth inhibitors released into the soil environment may be changed to other inhibitors by both bacteria and fungi, which metabolize the chemicals and then release them into the environment. Redrawn from S. B. Horsley, 1976. Allelopathic interference among plants, II. Physiological modes of action. In H. E. Wilcox and A. T. Hamer (Eds.), *Proceedings of the 4th North American Forest Biology Workshop.* Used with permission from U. S. Forest Service.

experiments do not tell much about the source of the allelochemicals, which may well come from degradation products of microbial metabolism. Futhermore, substances may be leached out of the roots that would not normally be released.

Release of volatile allelochemicals from plants has been most demonstrated for plants of arid and semiarid climates (Whittaker, 1971). A large number of plants are known to release volatiles inhibitory to associated plants and most of the chemicals identified are terpenoids (Weaver and Klarich, 1977).

The death of plants or plant parts and their contact with the soil results in the release and degradation of the organic matter with the consequent release of allelochemicals in significant quantities. In the process of dying, membrane permeability increases, which permits the release of water soluble inhibitors from within the cells. Figure 8.2 shows some inhibitors and their transformation products in the soil. Both bacteria and fungi may be involved in the transformations.

CLASSIFICATION OF ALLELOCHEMICALS

Based on their biological activity, allelochemicals can be classified as follows:

1. Phytotoxicants
2. Growth promoters

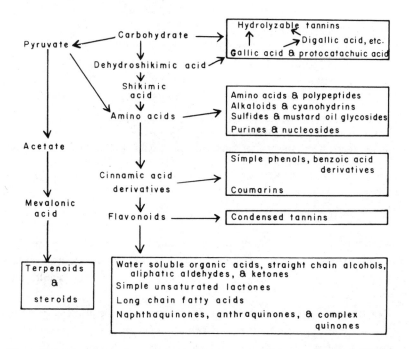

Figure 8.3. The metabolic origin of allelochemicals from the normal metabolism of plants is outlined. The allelochemicals are enclosed in the boxes. Redrawn from E. L. Rice, 1974. *Allelopathy*. Used with permission from Academic Press, New York.

3. Substrates for microorganisms

4. Predisposers for disease

5. Enhancers of root exudation

6. Agents for altering soil structure

In each of these categories they affect physiological processes in stressful ways that depend upon the species and environmental conditions.

The chemical nature of allelochemicals covers a wide range of classes of compounds and the list of compounds is extensive. Most of the allelochemicals arise from the activity of the mevalonic acid metabolic pathway (Figure 8.3) (Rice, 1974). Many of them are described as secondary metabolites with over 10,000 known (Swain, 1977), many of which serve as protective agents against competitors, predators, and pathogens.

ALLELOPATHY OCCURRENCE

How does one determine if allelopathy is occurring? Can allelopathy be separated from competition in an experimental procedure? Any methodology must prove that there is a cycling of a physiologically active substance or substances in an ecological system. Physiological activity is demonstrated by using a bioassay that reflects in a measurable way the presence of an allelochemical. The choice of an assay organism becomes important because one that is not sensitive to the allelochemical would not be advantageous. Chemical assays are often difficult because of the complexity of the systems and the low concentrations involved, sometimes over long periods of time.

What sorts of information are useful in determining allopathy? The concentration of the chemical in the environment and its effective threshold concentration should be known. Persistence in the environment is a factor; how long does the chemical last and is it continually produced over long periods of time. The physiological effects on receptor plants or organisms are useful in determining the role of the allelochemical in the system, as is knowledge of the uptake by other plants. Some allelochemicals are effective only on selected species so the selectivity in the system becomes important.

Tests for allelopathy must answer two questions: (1) Has it been demonstrated that the plant (donor) contains an inhibitory compound and is there a mechanism for its release and accumulation in the environment of the receptor plant? and (2) Have other factors such as competition been ruled out? It has been suggested that criteria similar to Koch's postulates be used as follows:

1. Identify symptoms; correlate interference with symptoms.

2. Rule out competition by determining that competition cannot account for all the observed effects.

3. Isolate and assay the chemical involved.

4. Simulate interference by supplying the chemical as it would occur in the ecosystem.

5. Monitor the release of the chemical for quantity and uptake.

6. If possible, correlate selectivity of the chemical with its role in the ecosystem.

There are a number of difficulties in carrying out all of these steps but progress is being made on the methodology. Symptoms of allelopathy are varied because of the varied nature of the compounds involved. Some of the more common ones are reduction in growth, reduced photosynthesis and respiration, impaired ability to absorb nutrients, chlorosis, deformation, death, and others that are more subtle.

PHYSIOLOGICAL ACTION OF ALLELOCHEMICALS

The inhibitory activity of allelochemicals is related to the amount absorbed and translocated to a site where physiological action of a toxic form has a detrimental effect. Some molecules of molecular weight as large as 600 are rapidly absorbed and translocated in the xylem and some molecules of molecular weight as large as 1400 to 1500 may be absorbed but not translocated. The usual processes of absorption have been implicated, that is, diffusion and active transport. In arid and semiarid regions the terpenoids have been implicated in inibitory reactions and in the temperate regions it is the phenolics such as benzoic and cinnamic acids and the coumarins. Inhibition of respiration and oxidative phosphorylation, photosynthetic carbon fixation, enzyme activity of various kinds, protein synthesis, and cell division and elongation have at one time or another been shown to be affected by allelochemicals (Horsley, 1976). It is evident that the allelochemicals cover such a broad range of compounds, that the effects on physiology of receptor plants is also varied and complex.

BREEDING FOR ECOLOGICAL ALLELOPATHIC ADVANTAGE

One of the ways in which knowledge of allelopathy might be used in agriculture is in the limitation of growth of unwanted plant species (weeds). It would be interesting if crop plants could be selected in a breeding program for their allelopathic capability. The same reasoning would apply to breeding for resistance to plant diseases and to insects.

Examples of the attempts to select for allelopathy against weeds are those of Lockerman and Putnam (1981) and of Fay and Duke (1977). Based on growth analysis of the effects of allelopathy of cucumber (*Cucumis sativus* L) accessions selected from the large breeding stock of the United States Department of Agriculture, Lockerman and Putnam concluded that there was an allelopathic suppression of growth of proso millet (*Panicum miliaceum* L) during the early stages of growth of the cucumber plants. The inhibitory effects disappeared 10 days after imbibition of the cucumber seeds. Was the early production of inhibitor by cucumber enough to give it an ecological advantage over weeds in a mixed population? The proso millet did not regain normal growth after the initial exposure to cucumber and data on net assimilation rates and relative growth rates showed that the accession cucumber was not competitively better than the nonallelopathic cultivar.

Some varieties of oats (*Avena sativum* L) produce roots that contain higher amounts of scopoletin than others. When oat plants were placed under drought or high temperature stress they produced as much as 25 times more scopoletin (Martin and Rademacher, 1960). Scopoletin was shown to inhibit root growth but recptor plants recovered from inhibition when scopoletin was removed from the nutrient solution (Fay and Duke, 1977). All of the 3000 accessions of oats produced fluorescent chemicals but in varying amounts. Only those that produced the most scopoletin were tested in the allelopathic experiments.

DISCUSSION QUESTIONS

1. In what forms does allelopathy occur among organisms?
2. Why is allelopathy considered an environmental stress?
3. The chemical interaction among plants has ecological significance. What are some of the possibilities?
4. In what ways is it possible to use the knowledge and principles involved in allelopathy to the advantage of agriculture?
5. What is the nature of the chemicals involved?

6. It is sometimes difficult to determine if there is a chemical interaction between one plant and another. What are some ways in which such a determination can be accomplished?

7. To be inhibitory, an allelochemical must effect a physiological process or processes. What are some of the ways in which allelochemicals bring about their effects?

8. Is breeding a practical way to bring about alleviation from allelopathic stress?

REFERENCES

Davis, E. F. 1928. The toxic principle of *Juglans nigra* as identified with synthetic juglone and its toxic effects on tomato and alfalfa plants. *Am. J. Bot.* 15:620.

del Moral, R., and C. H. Muller. 1969. Fog drip: A mechanism of toxin transport from *Eucalyptus globulus*. *Bull. Torrey Bot. Club* 96:467–475.

Fay, P. K., and W. B. Duke. 1977. An assessment of allelopathic potential in *Avena* germ plasm. *Weed Sci.* 25:224–228.

Hale, M. G., and L. D. Moore. 1979. Factors affecting root exudation II: 1970–1978. *Adv. Agron.* 31:93–123.

Horsley, S. B. 1976. Allelopathic interference among plants II. Physiological modes of action. In H. E. Wilcox and A. T. Hamer (Eds.), Proceedings of the 4th North American Forest Biol Workshop, College of Environmental Science and Forestry, Syracuse, New York. pp. 93–136.

Kaminsky, R. 1981. The microbial origin of allelopathic potential of *Adenostoma fasciculatum* H & A. *Ecol. Mon.* 51:365–382.

Leather, G. R. 1983. Sunflowers (*Helianthus annuus*) are allelopathic to weeds. *Weed Sci.* 31:37–42.

Lockerman, R. H., and A. R. Putnam. 1981. Mechanisms for differential interference among cucumber (*Cucumis sativus* L) accessions. *Bot. Gaz.* 142:427–430.

McWhorter, C. G. (Ed.). 1978. The role of secondary compounds in plant interactions (Allelopathy). USDA, ARS Research Planning Conference.

Martin, P., and B. Rademacher. 1960. Studies on the mutual influences of weeds and crops. *Br. Ecol. Soc. Symp.* 1:143–152.

Massey, A. B. 1925. Antagonism of the walnuts (*Juglans nigra* L and *J cinerea* L) in certain plant associations. *Phytopathology* 15:773–784.

Molisch, H. 1937. *Der Einfluss einer Pflanze auf die andere-Allelopathie.* Fischer, Jena.

Muller, C. H. 1966. The role of chemical inhibition (Allelopathy) in vegetational composition. *Bull. Torrey Bot. Club* 93:332-351.

Putnam, A. R., and W. B. Duke. 1978. Allelopathy in agro-ecosystems. *Annu. Rev. Phytopathol.* 16:431-451.

Rice, E. L. 1974. *Allelopathy.* Academic, New York.

Rice, E. L. 1979. Allelopathy—An update. *Bot. Rev.* 45:15-109.

Rovira, A. D. 1969. Plant root exudates. *Bot. Rev.* 35:35-59.

Schneiderman, F. J. 1927. The black walnut (*Juglans nigra* L) as a cause of the death of apple trees. *Phytopathology.* 17:519-540.

Swain, T. 1977. Secondary compounds as protective agents. *Annu. Rev. Plant Physiol.* 28:479-501.

Tukey, H. B., Jr. 1966. Leaching of metabolites from above ground plant parts and its implications. *Bull. Torrey Bot. Club* 93:385-401.

Tukey, H. B., Jr. 1969. Implications of allelopathy in agricultural plant science. *Bot. Rev.* 35:1-16.

Tukey, H. B., Jr. 1970. The leaching of substances from plants. *Annu. Rev. Plant Physiol.* 21:305-324.

Tukey, H. B., Jr., and R. A. Mecklenburg. 1964. Leaching of metabolites from foliage and subsequent reabsorption and redistribution of the leachate in plants. *Am. J. Bot.* 51:737-742.

Weaver, T. W., and D. Klarich. 1977. Allelopathic effects of volatile substances from *Artemisia tridentata* Nutt. *Am. Midland Nat.* 97:508-512.

Whittaker, R. H. 1971. The chemistry of communities. In U. S. Natl. Comm for IBP (Eds.), Biochemical interactions among plants. Natl. Acad. Sci., Washington, DC, pp. 10-18.

Whittaker, R. H., and P. P. Feeney. 1971. Allelochemics: Chemical interaction between species. *Science (Washington)* 171:757-770.

9

EFFECTS OF STRESS
ON MEMBRANES

Membranes are probably the first line of defense against adverse environmental changes. They act as sensors of the environment and initiate internal changes that lead to metabolic responses of a nature that may render the plant tolerant of adverse alterations in the environment. It is necessary, therefore, to have a knowledge of the structure and function of membranes and of how they react to stress.

MEMBRANE STRUCTURE AND FUNCTION

Membranes are integral parts of cellular organelles such as chloroplasts, mitochondria, vacuoles, nuclei, golgi bodies, and endoplasmic reticula. The cytoplasm is surrounded by a single unit membrane called the plasmalemma but some organelles, such as chloroplasts and mitochondria, have a double membrane system. Membranes are composed of lipids and proteins organized in a bilayer configuration as illustrated in Figure 9.1. The percentage of lipid and protein varies with the organelle and with the plant

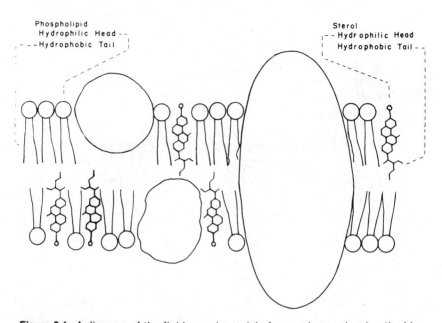

Figure 9.1. A diagram of the fluid mosaic model of a membrane showing the bilayer nature of structure. The large bodies are proteins that may traverse across the membrane or lie in either layer of lipid. The hydrophilic heads of both the lipids and the sterols are oriented toward the surface of the membrane and the hydrophobic tails toward the interior.

species. In addition, the composition of the outer and the inner sides of the bilayer may differ within the same organelle. The predominant lipid components of the bilayer are phospholipids. However, some organelles have an abundance of other types of lipids such as galactolipids.

Phospholipids consist of a glycerol molecule to which are attached two fatty acyl groups that may differ in carbon chain length and degree of unsaturation. The third position of the glycerol molecule is occupied by a phosphate group that can attach to various carbohydrates, amines, and amino acids. The fatty acyl groups are oriented toward the center of the bilayer so that acyl groups from each bilayer are oriented toward each other in the center of the bilayer. The more polar phosphate heads are oriented toward the outer surfaces of the membrane.

Proteins are also quite variable in their structure and location in the membrane. For example, they can be associated with the membrane sur-

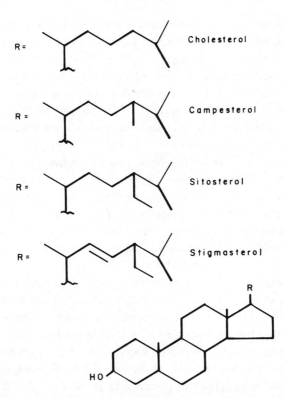

Figure 9.2. The structure of the four common phytosterols. The basic molecule is the structure at the bottom of the diagram. The sterols differ only in the side chain structure depicted as R. From G. Grunwald, 1975. Plant sterols. *Annu. Rev. Plant Physiol.* 26:209–236. Used with permission from Annual Reviews Inc., Palo Alto, CA.

face, embedded partially or totally in the membrane, or traverse the entire thickness of the membrane.

Components called sterols are also associated with membrane structure. Cholesterol, a common constituent of animal membranes, is also found in plant membranes; however, other sterols such as sitosterol, stigmasterol, and campesterol, are more abundant (Figure 9.2). The major plant sterols differ from cholesterol in that they may have additional methyl or ethyl groups attached at carbon-24 of the side chain. In addition, differences in steric configuration and degree of unsaturation of the side chain can occur in different plant species. Sterols are heterocyclic ring structures (four rings) with a hydrocarbon side chain attached to ring IV. Ring I has a hydroxyl group attached and the molecule is inserted into the bilayer of the membrane between the phospholipids so that the hydrocarbon portion of the molecule is oriented toward the center of the bilayer and the hydroxyl group toward the polar end groups of the phospholipids. Sterols have the function of stabilizing the phospholipids in the membranes, making them less fluid and thus less permeable. Not all membranes have sterols, however.

Membranes are dynamic systems and are constantly in a state of motion. The fluidity of the membrane depends on the chemical structure of the lipids and proteins that comprise the membrane as well as the way they are organized within the membrane. Environmental influences such as temperature, light, ion interaction, degree of hydration, and hormonal effects, can influence membrane fluidity. Generally, the more fluid a membrane is, the more permeable it will be.

The acyl groups of the phospholipids have a great deal to do with the fluid nature of the membrane. The longer and more unsaturated the fatty acyl groups are, the more fluid the membrane will be. The positional relationship of the phospholipids with respect to the proteins in the membrane will also affect the fluidity of the membrane. If the phospholipids are tightly held to a protein inserted in the membrane, the fluid nature of the membrane in the vicinity of the protein may be reduced. Also, if sterols are a part of the membrane structure, they tend to make membranes less fluid and consequently less permeable. In animals there seems to be a relationship between the degree of unsaturation of fatty acyl groups of the phospoholipids and the presence or abscence of sterols in membranes. The more highly unsaturated the fatty acyl groups are, the less likely that sterol will be found in the membrane.

The fluid nature of the membrane is affected by temperature. Increases or decreases in temperature can influence the kinetic motion of the membrane molecules, thus changing the fluidity and permeability of the membrane. Temperature can also modify the chemical structure of the fatty acyl groups of the phospholipids. Elevated temperatures tend to increase the degree of saturation in the membrane lipids, while reduced temperatures result in a greater degree of unsaturation. In the former, the tendency would be for an increase in rigidity of the acyl groups, while in the latter a more flexible and more fluid condition would exist. Such changes in composition of the fats may moderate the effects of temperature on the kinetic motion of the membrane molecules in response to changing temperature.

Components can change position in the membrane or be totally lost. Movement within the membrane bilayer is a common occurrence and is influenced by the kinetic movement of membrane molecules. Occasionally a molecule may flip-flop and be positioned on the opposite side of the membrane. However, this occurs infrequently. Lipids with very hydrophilic heads may be lost from the membrane into the cytosol of the cell. This happens with greater frequency than the flip-flop phenomenon, occurring about every 3 h. Components can bob in and out of the membrane by extending above or below the surface of the bilayer. This occurs when charge characteristics of either phospholipid heads or proteins change and create different affinities among the bilayer components and/or the external environment of the membrane.

Membranes have multiple functions in plants. One of the most obvious is related to their semipermeable nature, which determines what can enter and/or leave the cell or organelle. Several factors control the ability of a molecule or an ion to penetrate a membrane. Size, electrical charge, and chemical compatability with the membrane are important considerations. Generally, water has very little difficulty in crossing membranes. However, the rate of penetration may be governed by the fluidity of the membrane and also the kinetic energy associated with the water molecule. Charged particles, such as ions, have great difficulty in crossing membranes by simple diffusion. Thus, membrane carriers are often required for ion transport across the bilayer.

The lipids and proteins that comprise the membrane structure have differing polarities. For example, phospholipids have both a polar and a nonpolar portion. The phosphate head of the molecule will often be charged and is polar, while the fatty acyl group is not charged and is nonpolar. Compounds exhibiting both characteristics are termed amphipathic molecules. Proteins are usually highly charged and so they are very polar molecules.

The orientation of the proteins and phospholipids in the membrane is related to their polar and nonpolar nature. The polar nature of the phospholipid heads makes them hydrophilic, causing them to be aligned in an aqueous environment. Conversely, the hydrophobic nature of the fatty acyl groups causes this portion of the phospholipid molecule to align toward the center of the membrane.

The polar and nonpolar nature of membrane components also controls the type of molecule that will pass through a membrane. Substances with polarity characteristics similar to those of the membrane will have less difficulty penetrating the bilayer.

Transport across membranes occurs by passive and active mechanisms. Passive transport requires no metabolic energy and is a result of differences in concentration and/or electrical potential gradients across the membrane. In the latter case, diffusion of ionic substances is involved in which different charged particles diffuse at different rates, creating a gradient of electrical potential along the path of diffusion. Concentration also contributes to the movement of the ions thus, acting in conjunction with electrical potential to facilitate diffusion. Diffusion will continue to occur across a membrane until an equilibrium of molecules is established or an electrochemical equilibrium is achieved.

Facilitated diffusion also occurs as a means of transport across biological membranes. In this instance, specific proteins called permeases or carriers are involved in transport. The mechanism of transport is largely unknown. However, there are specific permeases known for the transport of glucose, maltose, and lactose in some bacteria. The proteins of the permeases are very specific for their attachment to particular substrates. Transport may involve conformational changes in the protein resulting in the attached molecule being physically moved into the membrane where it is chemically altered to the extent that it cannot move back across the membrane, or its solubility characteristics become more compatible with those of the inner portion of the membrane. Several types of permeases have been characterized in membranes. Uniports are permeases responsible for the transport of only one substance in one direction, antiport permeases transport two different substances in opposite directions simultanelously, and symport permeases transport two different molecules in the same direction simultaneously.

Active transport can occur against a concentration gradient resulting in the accumulation of higher concentrations within the cell than occur in the external environment. Several types of active transport systems are known to occur in biological systems. Some are dependent on proton gradients established by metabolic systems such as respiration and photosynthesis, some on ionic gradients, and some on ATP (adenosine triphosphate) operating specific ATPases.

In all instances, transport is dependent on charge separation provided by the insulating effects of membranes. Proteins serve as carriers for the active transport of some ions and molecules such as simple sugars and amino acids. In addition, they have enzymatic functions in numerous biochemical reactions that may occur on the membrane surface or within the bilayer. The electron transport reactions of respiration and the light reactions of photosynthesis that occur in the membranes of the mitochondria and chloroplast thylakoids, respectively, are two examples in which membrane proteins play an important role in metabolic functions. Here, membranes serve as charge insulators when charge separation is critical to metabolic events, such as respiration, in which electrons and protons are generated in oxidation–reduction reactions. The reduction of NADH + H^+ liberates two hydrogen ions to the exterior of the membrane and the electrons are passed through an electron transfer system within the membrane until they finally unite with oxygen and hydrogen at the inner surface. The hydrogen ions extruded to the exterior of the membrane then serve as a catalyst to activate the enzyme ATPase, which then forms ATP or accumulates until it can function in the cotransport of other ions or molecules. Proton generation can also occur as a result of hydrolysis of ATP.

The insulating feature of membranes in regard to their ability to maintain charge separation, as occurs in respiration and photosynthesis, is not only important for the generation of ATP, but also represents an important mechanism for the transport of ions and other charged molecules into and out of organelles. Such redistribution of ions is facilitated through changes in electrical potential differences across membranes, which will tend to equilibrate through the transport of ions of opposite charge.

TEMPERATURE AND MEMBRANE FUNCTION

Temperature is one of the most critical environmental parameters known to affect membrane structure and function. The degree of fluidity of the cell membrane greatly affects the ability of cells to survive and function. In general, membrane fluidity increases with a rise in external temperature and decreases with a lowering of temperature. However, membrane structure and composition determine the extent to which temperature affects the fluid nature of the membrane.

The phospholipids that comprise the lipid component of membranes are particularly sensitive to temperature change. The physical configuration of the fatty acyl groups of these molecules changes when exposed to differing temperatures. Low temperatures cause a more structured configuration, making the membrane more rigid. At elevated temperatures the acyl groups are less organized and more randomly configured and the membrane takes on a more fluid nature. Phospholipids have a specific temperature at which they are transformed from the fluid state (liquid-crystalline lattice) to a more solid or gel state. The temperature at which this change occurs is called the transition temperature (Tc).

In biological membranes the lipids are a mixture of different phospholipids and the Tc occurs over a broad temperature range. The carbon length of the fatty acyl groups and the number of double bonds in the carbon chains also affects the Tc. Longer chain lengths and fewer double bonds in the chains tend to raise the temperature at which the transition to the gel state occurs. The opposite is true when the chain lengths are shortened and are more unsaturated (more double bonds). Some phospholipids also have fatty acyl groups that are not straight chained but are branched. These types of fatty acids tend to extend the Tc to a lower level.

Also, the insertion of sterols, such as cholesterol, between the phospholipids tends to make the phospholipid acyl groups more rigid above the Tc of the lipid and to create a greater thermal motion below the Tc. Equimolar concentrations of sterol/phospholipid completely stop any transition to a crystalline gel state and in some instances abolish sharp increases in water permeability that occur at the Tc. Cell growth occurs at or above the membrane phase transition temperature and is restricted or abolished below it.

Since biomembranes are composed of phospholipid mixtures, they often undergo lateral phase separation rather than simple phase transition as a result of reduced temperatures. Phase separation occurs when phospholipid molecules are sufficiently heterogeneous that, at reduced temperatures, they aggregate in more homologous groups creating islands of molecules that sometimes isolate and relocate proteins in a lateral movement. Such aggregations of components could result in the formation of fractures in the membrane leading to leakage of cellular components.

One of the responses to chilling temperatures in plants is the production of increased levels of unsaturated fatty acids. By increasing the degree of unsaturation, the fluidity, and functionality of the fatty acyl groups of the membrane phospholipids is maintained. It appears that different membranes within the same cell respond to temperatures at different rates suggesting different mechanisms for acclimating to changes in temperature. One hypothesis suggests that rapid increases in unsaturated fats in membranes is the result of the activation of desaturase enzymes embedded in the membrane. These enzymes are activated as a result of the initial increase in the rigidity of the phospholipid acyl groups, which causes a repositioning of the enzyme so that it is next to its substrate, or, the configuration of the enzyme may be physically altered, resulting in activation.

The slower desaturation response observed in some membranes is thought to be related to synthesis of new lipids and transport to the site of incorporation in the membrane. Thus, the rapid reaction of desaturases in controlling membrane fluidity represents an immediate response to environmental stimuli that affects the physical nature of membrane components. Such responses have obvious adaptive advantages for the organism in a changing environment.

The insertion of sterols into the membrane represents another means by which membrane fluidity can be controlled. Sterols decrease membrane fluidity above the Tc and increase fluidity below the Tc. Thus, those factors controlling sterol metabolism in plants could have a controlling effect on membrane fluidity. Sterol metabolism is responsive to environmental parameters such as light intensity, temperature, and changes in response to red-far-red light (R-FR) interconversions of phytochrome.

Higher plants have several different sterol components that are known to fluctuate in relative composition and total amounts with stage of development. This is in contrast to higher animals, which generally have but one sterol, cholesterol, as the single predominant sterol component in their tissues. Such differences in sterol complexity in higher plants compared to animals has raised questions as to their potential function. In view of the stationary nature of plants and their inability to control their internal tissue temperatures as warm blooded animals do, such a diversity may represent a molecularly diverse approach at maintaining membrane fluidity under differing environmental influences. The major difference between most higher plant sterols and chloesterol is a greater number of carbon atoms (C-28 and C-29) and a higher degree of unsaturation in the side chain of the sterol molecule. Such differences however, could be very functional in the control of membrane fluidity under varying environmental conditions.

IONIC INTERACTIONS AND MEMBRANE FUNCTION

Divalent and monvalent cations have a considerable influence on the fluid nature of biomembranes. Divalent cations are generally more effective that monovalent cations. Ca^{2+} and Mg^{2+} are two divalent cations that have been studied extensively using both artificial and biomembrane systems. How these ions interact with membranes depends on the type and abundance of membrane components present, as well as how they are arranged in the bilayer. The phospholipids appear to be the most drastically affected components of the membrane. However, interactions with proteins are also known to occur.

Ca^{2+} is the most effective divalent cation relative to changing membrane fluidity. It reacts primarily with negatively charged or acidic phospholipid head groups, causing the Tc of the membrane to shift toward higher temperatures. Thus, Ca^{2+} has a condensing effect on the phospholipids in the bilayer, making them more rigid or gel-like. Ca^{2+} not only has an effect on the head groups of the phospholipids but in acidic phospholipids, such as phosphatidylserine, the Tc is not as drastically affected when the fatty acyl groups are unsaturated as when they are saturated. Neutral phospholipids, such as phosphatidylcholine, also bind to Ca^{2+} and Mg^{2+}; however, neither element is as effective as when bound to acidic phospholipids.

Mixtures of neutral and acidic phospholipids often result in lateral phase transition of lipid components resulting in domains of neutral and acidic phospholipids in the membrane. Although acid phospholipids may be a minority in membranes, the formation of such domains around enzymes may have considerable effect on their activity. Also, changing the ratio of acidic/neutral phospholipids in specific areas of the membrane could affect the fluidity and permeability characteristics locally in the membrane. Mg^{2+} does not induce lateral phase transition. However, it does reduce the effectiveness of Ca^{2+} when applied in combination. The rate and extent to which Ca^{2+} induces domain formation in membranes is related to Ca^{2+} concentration, time of exposure, and the ratio of neutral to acidic phospholipids. In addition Ca^{2+} replaces the water of hydration associated with the phospholipid heads and at high enough concentrations water is completely displaced. This in turn may have a direct effect on enzyme configuration and function, thus contributing to further alteration in membrane fluidity. Mg^{2+} is not as effective as Ca^{2+} in this regard.

Exactly how Ca^{2+} and Mg^{2+} interact with the phospholipids of the membrane is not completely clear. However, it is known that a reduction in negative surface charge density occurs, which could affect electrical potential differences across membranes. Divalent cations also affect hydrogen bonding, water of hydration, and packing of polar head groups. In addition, it has been shown that Ca^{2+} can bind with the hydroxyl group of cholesterol so that perhaps a Ca^{2+}-to-sterol interaction is part of the mechanism for a "tighter" packing of phospholipid polar head groups.

Several physiological disorders are known to exist in plants as a result of Ca^{2+} deficiency. These include water core of apple, blossom end rot of tomatoes, tulip topple, and leaf and hypocotyl necrosis. All of these disorders have similar symptoms of water-soaked tissues characteristic of loss of membrane integrity and increased leakage.

Changes in the pH in the vicinity of the membrane surface also affect the fluidity of the membrane. Hydrogen bonding between molecules of the membrane is important in maintaining membrane integrity. High pH results in complete extraction of protons resulting in a drop in the Tc. The reverse is true when pH is lowered. For a given phospholipid molecule three states of the bilayer membrane can be recognized: fully protonated, partially protonated, and deprotonated. Fully protonated and deprotonated systems differ in Tc by about 6°C for a single proton system such as phosphatidylglycerol and about 12°C for a two proton system such as phosphatidic acid. Tc is higher for the fully protonated state as compared to the deprotonated state. The partially deprotonated state represents a Tc temperature far above the other states of protonation. This difference may be explained by greater stabilization of hydrogen bonding. Protonation can also replace Ca^{2+} and can by itself induce lateral phase transition, which can be reversed by raising the pH in a buffered salt solution containing monovalent cations. Monovalent cations seldom bind negatively charged phospholipids but greatly affect proton binding. This can influence the phase transition of negatively charged phospholipids by interacting with partially protonated systems causing an increased dissociation of the protons and a decrease in the Tc. Monovalent cations can interact in the deprotonated state at low salt concentrations to slightly increase Tc or at high salt concentrations to strongly increase Tc.

Potassium is a very important monovalent cation in plants that helps to maintain and regulate cell turgor under stress conditions. It also is known to stimulate ATPase activity in membranes that may be involved in active transport of other ions. Several so-called "potassium pump" systems are known to exist in plants where K^+ exchanges for other ions to maintain osmotic or ionic balance. Such functions of K^+ in plants are not inconsistant with the effects of monovalent cations in membranes in general.

MEMBRANES AND DEHYDRATION STRESS

Water is a critical component for the stabilization of membrane structure. When in an aqueous environment in excess of 20 to 30% water, the amphipoteric nature of the phospholipid molecules (polar hydrophilic head, nonpolar hydrophobic acyl groups) causes them to form the most thermodynamically stable configuration typical of the bilayer organization of biomembranes. That is, the nonpolar acyl groups align themselves toward each other in the center of the bilayer with the polar phospholipid heads oriented toward the aqueous environment. Removal of water from the membrane to the extent that the percentage of water is below the critical 20 to 30% level causes a rearrangement of the phospholipids in a hexagonal configuration that is hydrophobic and is pierced by long water-filled channels lined by the polar heads of the phospholipids.

Since the formation of lipid droplets has been observed frequently in cells under desiccation stress, it has been concluded that membrane lipids also may be ejected from membranes. Such droplets disappear upon rehydration.

The structure of the proteins in the membranes seems to be little affected by dehydration; however, proteins may be displaced in the membrane, or totally excluded, as a result of lipid rearrangements. The rate, severity of dehydration, and composition of the membrane are all implicated in the degree of damage sustained by a membrane as a result of dehydration. Upon rehydration, cellular membranes are reconstituted. The rate and degree of restoration of function depends on the amount of damage sustained by individual membranes. The composition and structure of some types of membranes may make them either more tolerant or more susceptible to dehydration stress and the rate of reconstitution and reestablishment of functionality of membranes may be the basic difference between drought-tolerant and drought-sensitive plants. The rate of rehydration may also affect physiological and metabolic functions, particularly if rehydration is rapid and reconstitution of cellular membranes is slow.

Such a situation could result in the leaching of ions, of metabolic substrates, and of other essential components necessary for the reestablishment of membrane structure.

The ability to reestablish membrane function of mitochondria and chloroplasts that have been dehydrated is of paramount importance because the electron transport functions and light-capturing mechanisms are localized within the membrane structure of such organelles. If essential proteins required in such reactions are ejected or rearranged in the membrane, the energy relations of the cell will be severely affected until such components can be resynthesized or reestablished.

The concentration of ions in cells as a result of dehydration could also affect membrane structure and function. Potassium is a very abundant ion in most plant tissues. Accumulation of this ion has been known to displace Ca^{2+} from the membrane, resulting in destabilization.

LIGHT AND MEMBRANE PERMEABILITY

Light, through its action on phytochrome, also may have an effect on membrane structure and permeability. Although actual proof of phytochrome occurring in membranes is lacking, considerable evidence suggests that it has a controlling influence on membrane permeability, ion transport, and the transport of metabolites such as phytohormones. Phytochrome responses in plants can be arbitrarily divided into three categories: those that occur rapidly (0 to 15 min) and are considered membrane associated responses; those that are expressed in 1 to 3 h, which involve changes in amounts or activity of specific enzymes and may involve changes in gene expression; and those that are much slower and are expressed as developmental responses. It is conceivable that the primary response leading to enzyme expression and ultimately plant development is related to phytochrome mediated changes in membrane structure and permeability. Since phytochrome has been described as being an elongated nonglobular protein, it is possible the R-FR light-mediated changes in configuration of the phytochrome protein could change it from a peripheral position on the surface of the membrane to a transmembrane position. Such a shift would create aqueous pores in the membrane through which nonspecific transport of substances across the membrane could occur.

Pfr may be the form of phytochrome that transcends the membrane and does so by complexing with membrane sterols creating native channel-forming quasi-ionophores. In addition, since several studies have indicated rapid increases in ATP and reductions in ADP after red-light exposure, nonspecific transport could possibly be modulated through energy metabolism at the membrane surface. Thus, active transport mechanisms may be altered and transport and functionality of phytohormones may also be affected through phytochrome alteration of membrane structure.

MEMBRANE PERMEABILITY AND PHYTOHORMONES

Virtually all the phytohormones are known to affect membrane permeability. In some instances they may combine with protein receptors in the membrane resulting in release of effectors or factors that affect transcription processes. In this way developmental processes may be changed. Such an interaction, for example, has been demonstrated with the use of IAA.

Treatment of plants with phytohormones has often been observed to affect the uptake and translocation of ions and nutrients. Indirectly such changes suggest interaction with cell membranes. The relationship of ABA and CK in controlling stomatal movement, K^+ fluxes in the guard cells, and water transport are suggestive of membrane permeability and transport control.

In some instances the composition of the phopholipids of membranes may have an effect on the interaction of phytohormones relative to membrane permeability control. For example, IAA changes the physical characteristics of synthetic membrane systems depending on the phospholipid composition. Synthetic auxins have also been shown to result in the thinning of membranes when applied to living cells, which suggests a physical change in the membrane structure. Some phytohormones may compete with certain membrane components for sites in the membrane that could directly affect membrane structure and physical characteristics. Such a function has been attributed to some gibberellins in regard to replacing sterols in membranes. The general structure of gibberellins is very similar to that of sterols and gibberellins have been used to increase permeability in synthetic membrane systems.

DISCUSSION QUESTIONS

1. What are biological membranes composed of chemically?

2. Describe how the chemical components of biological membranes are organized within the membrane.

3. What are the functions of sterols in membranes?

4. What is meant by membrane fluidity? How is fluidity controlled through changes in the chemical composition of membrane components?

5. Why is fluidity important in biological membranes?

6. What is the term used to describe a molecule that exhibits both polar and nonpolar characteristics?

7. Define facilitated diffusion.

8. List and define the various types of permeases.

9. Name three types of active transport systems.

10. Describe how temperature affects membrane fluidity and indicate how plant membranes may adjust to maintain a fluid nature.

11. What is meant by the transition temperature (Tc) of a membrane?

12. Define lateral phase separation as it pertains to membranes.

13. How might desaturase enzymes function in membranes to quickly regulate membrane fluidity under changing environmental conditions?

14. How does Ca^{2+} interact with membranes to affect fluidity.

15. How does pH affect the Tc of biological membranes?

16. Describe the effects of dehydration stress on membrane structure and composition.

17. What factors affect the degree of damage sustained by a membrane as a result of dehydration? How?

18. What are the three categories of phytochrome responses in plants?

19. How might phytochrome be involved with membrane permeability?

20. What relationship do phytohormones have to membrane permeability?

REFERENCES

Chapman, D. 1983. Biomembrane fluidity: The concept and its development. In R. C. Aloia (Ed.), *Membrane Fluidity in Biology*, Vol. 2. Academic, New York, pp. 5-42.

Duzgunes, N., and D. Papahadjopoulos. 1983. Ionotropic effects on phospholipid membranes: Calcium-magnesium specificity in binding, fluidity and fusion. In R. C. Aloia (Ed.), *Membrane Fluidity in Biology*, Vol. 2. Academic, New York, pp. 187-216.

Eibl, H. 1983. The effect of proton and of monovalent cations on membrane fluidity. In R. C. Aloia (Ed.), *Membrane Fluidity in Biology*, Vol. 2. Academic, New York, pp. 217-236.

Gibbons, G. F., K. A. Mitropoulos, and N. B. Myant. 1982. *Biochemistry of Cholesterol*. Elsevier Biomedical, New York, pp. 303-342.

Grunwald, C. 1975. Plant sterols. *Annu. Rev. Plant Physiol.* 26:209-236.

Pringle, M. J., and D. Chapman. 1981. Biomembrane structure and effects of temperature. In G. J. Morris and J. A. Clarke (Eds.), *Effects of Low Temperatures on Biological Membranes*. Academic, New York, pp. 21-37.

Quail, P. H. 1983. Rapid action of phytochrome in photomorphogenesis. In W. Shropshire, Jr., and H. Mohr (Eds.), *Photomorphogenesis*, Vol. 16A. Springer-Verlag, New York, pp. 178-212.

Robertson, R. N. 1983. *The Lively Membranes*. Cambridge Univ. Press, New York.

Roth-Bejerano, N., and R. E. Kendrick. 1979. Effects of filipin and steroids on phytochrome pelletability. *Plant Physiol.* 63:503-506.

Smith, H.. 1976. The mechanism of action and function of phytochrome. In H. Smith (Ed.), *Light and Plant Development*. Butterworths, London, pp. 493-502.

10

THE ROLE OF PHYTOHORMONES IN STRESSED PLANTS

Many reactions of plants to environmental stresses involve morphogenetic changes, which implies that stress causes changes in phytohormonal balance. Are alterations of phytohormone balance primary or secondary reactions? Can reactions to stress be altered by applying phytohormone-like substances? Research approaches to obtain answers to these questions involve (1) measurements of phytohormone concentrations of stressed and nonstressed plants and of susceptible and resistant plants; (2) manipulations of phytohormone levels to determine if plant reactions to stress are affected; and (3) trying to cause stress symptoms by applying phytohormones or regulators to plants and measuring changes in reaction to stress. In the discussion that follows remember these points as the facts and examples are presented.

PHYTOHORMONE RESPONSE AND WATER RELATIONS

Water stress affects hormone balances, which in turn control plant developmental patterns. Virtually all of the phytohormones are affected by water stress but most available information supports the hypothesis that abscisic acid (ABA), cytokinins (CK), and ethylene (ETH) are most

145

involved in interactions that control water balance, while indoleacetic acid (IAA) and the gibberellins (GA) are involved to a lesser degree.

Indoleacetic Acid

The concentration of IAA and its polar translocation are reduced when plants are under water stress. Plum and apple trees have lower IAA levels during drought than after a rainy period. The reduced concentration may result from an increase in concentration of IAA oxidase. For example, levels of the enzyme have been shown to increase in a linear fashion in tomato as water potential decreases from -0.2 to -1.5 MPa. Further evidence that auxins play a role in drought tolerance comes from the results of experiments that show that exogenous applications of IAA or 2,4-dichlorophenoxyacetic acid (2,4-D), can overcome drought-induced growth inhibition of wheat plants when the drought coincides with seed head formation, flowering, or the milk stage of seed development. Water deficits have also been linked to the loss of polar transport of IAA in pea stems.

Gibberellin

A decrease in concentration of GA occurs when plants are exposed to a water deficit or waterlogging conditions. For example, lettuce leaves exposed to water stress exhibit a lowered GA concentration, which returns to its previous level when the plants are released from the stress.

Because GA synthesis occurs in root tips, waterlogging causes a reduction in synthesis that is reflected in the amount translocated to the shoots. Such reductions in concentration can be measured in the xylem sap of waterlogged tomato plants. If exogenous GA is applied to the apical bud of tomatoes within 7 days after flooding, the inhibitory effects of waterlogging are overcome.

Ethylene

Water stress affects the concentration of ethylene (ETH) by increasing its synthesis. Ethylene is synthesized from aminocyclopropanecarboxylate (ACC), which arises from S-adenosylmethionine (SAM). Water stress, by increasing the synthesis of ACC, can cause an increase in ETH. Induction of ETH production by water stress has been demonstrated in a variety of

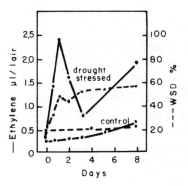

Figure 10.1. Relationship of water saturation deficit (WSD) and ethylene generation in broad bean leaves. Redrawn from S. T. C. Wright, 1978. Phytohormones and stress phenomena. In D. S. Letham, P. B. Goodwin, and T. J. V. Higgins (Eds.), *Phytohormones and Related Compounds—A Comprehensive Treatise.* Used with permission from Elsevier Science Publ. B.V., Amsterdam.

plants. When the water potential of cotton petioles approached -2.0 MPa, ETH concentration increased. In wheat tissue, water loss of 10% initiated an increase and in orange leaves an increase was observed 2 h after the initiation of water stress. An increase in ETH produced by broad bean occurred when the water saturation deficit (WSD) increased from 20 to 50% (Figure 10.1). Since exogenously applied ABA can also induce ETH production, increase in ETH caused by water deficits may be an indirect effect as a result of an increase in ABA concentrations within tissues. Such reasoning is further supported by the fact that ETH exerts a controlling influence on leaf abscission and abscission of young fruits in cotton plants. Older leaves, which have a higher concentration of ABA, respond more readily to increases in ETH levels than younger leaves. Furthermore, treatment of cotton leaves with ETH raises the threshold water potential required to induce abscission from -1.7 to -0.7 MPa. Increases in ETH levels of cotton bolls subjected to water deficits result in abscission. On the other hand, ETH may not cause an increase in concentration of ABA because the increase occurs immediately after application and at a faster rate than the increase in concentration of ETH. There is no evidence that ABA is involved in water-stress-induced cotton boll abscission even though the process is believed to be mediated by ETH.

The increase in ETH induced by water deficits may also have an effect on the translocation of IAA. In cotton, lowering the water potential from -0.8 to -1.2 MPa reduced IAA translocation by 50% and similar effects have been observed in fir trees, *Abies balsamea.* Such observations support the idea that a possible initial step in drought-stress-induced abscission may be inhibition of IAA translocation.

In addition to water deficits, the stress imposed by waterlogging of the soil also induces ETH synthesis. If ACC accumulates in the roots subjected to the anaerobic conditions caused by flooding and ACC is subsequently translocated to the shoots, it may be converted to ETH in the presence of oxygen in the shoot environment. When subjected to flooded conditions, an increase in ETH in shoots has been observed in a number of crop species such as dwarf bean, broad bean, chrysanthemum, maize, radish, sunflower, and tomato, but not in rice or barley.

Cytokinins

Because CK are synthesized in the roots and subsequently translocated to the shoots in the xylem, effects of drought stress on CK may be caused by the transmission of a signal of stress from the shoots to the roots. According to some investigators such a signal could be brought about by the translocation of ABA to the roots or by transmission of an osmotic signal. One example of reduction in CK synthesis in the roots as a result of a water deficit occurs in tobacco in which the reduction occurs as quickly as 30 min after initiation of the stress (Table 10.1).

Leaf senescence is a common result of water deficits so that a reduction in CK has been implicated. Application of exogenous CK to water-stressed plants can prevent senescence but CK also play a role in the control of stomatal aperture and the associated leaf water potential. The dual roles of CK in preventing protein loss from leaves and in stomatal aperture control

TABLE 10.1   Cytokinin Activity in Leaves and Exudates of Tobacco Plants

	Callus weight (mg)	Kinetin equivalent (μg/g FW)	Activity (%)
Exudate			
Nonstressed	236.2	0.140	100
Stressed	107.5	0.055	39
Leaves			
Nonstressed	115.8	0.044	100
Stressed	87.7	0.026	59

Source. C. Itai and Y. Vaadia. 1971. Cytokinin activity in water stressed shoots. *Plant Physiol.* 47:87–90. Used with permission from Plant Physiology, Waltham, MA.

TABLE 10.2   **The Effect of Wilting on the ABA Levels in Various Plant Organs**

Plant Organ	μg equiv ABA/kg FW	
	Controls	Wilted
Avocado (*Persea gratissima*)		
Roots	27	85
Sunflower (*Helianthus* sp)		
Lateral roots	13	19
Castor oil plant (*Ricinus communis* cv. Bigsonii)		
Leaves	31	410
Apices	131	830
Dwarf bean seedlings (*Phaseolus vulgaris* cv. Canadian Wonder)		
Leaves	19	335
Apices	78	827
Stems	23	288
Roots	3	160
Immature pods	74	609
Immature seeds	7	331
Datura sanguinea flowers	54	511
Cestrum newellii flowers	48	385
Wheat (*Triticum aestivum* cv. Hybrid 46)		
Etiolated leaves	28	294
Green leaves	44	257

Source. S. T. C. Wright. 1978. Phytohormones and stress phenomena. In D. S. Letham, P. B. Goodwin, and T. J. V. Higgins (Eds.), *Phytohormones and Related Compounds—A Comprehensive Treatise*. Used with permission from Elsevier Science Publishers, B. V., Amsterdam.

represent a complex interrelationship in cause and effect reactions that are not clearly understood. An interaction with the effects of ABA is suspected, as outlined in the next section.

Abscisic Acid

It has been well documented that water deficits cause an increase in ABA in a number of species of higher plants (Table 10.2). Accumulation occurs primarily in the leaves but essentially all organs show some increase. An increase in the concentration of ABA in tissues can be the result of release from bound forms, an increase in the rate of synthesis, a decrease in the rate of destruction, or a decrease in the rate of movement out of the tissue. Synthesis of ABA may be triggered by the reduction in turgor and the accompanying reduction in water potential. A search for a threshold water potential below which ABA synthesis increases and above which there is none, has had little success.

Under conditions of drought, a higher and stable level of ABA occurs because of the interchange among the various forms of ABA, such as conjugates and metabolites like phaseic acid. If a drought stress lasts for a comparatively long time the levels of ABA will decrease. Upon rehydration and release from drought stress the ABA levels decrease but at a slower rate than the initial rate of increase. The slow recovery rates give plants an adaptive advantage under conditions of intermittent water deficits.

Probably the most investigated role of ABA in relation to drought stress is its involvement in the control of stomatal aperture. ABA can initiate the closing or inhibit the opening of stomates and this role is illustrated most clearly in the mutant tomato, flacca, in which no ABA accumulates when the plant is exposed to water deficits. The stomates stay open despite loss of turgor and wilt when exposed to light. Closure occurs only when ABA is applied to the leaves.

The response to ABA is not always as clear as it is with flacca and stomatal aperture does not always reflect ABA levels. Such deviations may reflect the compartmentalization of ABA in organelles of the leaf cells. Let us consider the following scenario for some mesophytic plants that have been studied. The initial response of the plants to drought is a decrease in the leaf water potential, which results in the release of ABA from mesophyll chloroplasts followed by a rapid synthesis of ABA. ABA may also be supplied through transpiration-induced movement of water to the leaves. If the transpiration stream is terminated, the stomata open even though ABA is still in high concentration in the leaf. The explanation given has been that the ABA in the leaf exists in compartments and may not be available to the guard cells. The effects of previous stress conditions also may alter subsequent reactions of the plant to water potential changes of the environment but plants differ in their response. Some are more sensitive to stress as a result of previous exposure and some show a delayed response to stress after a period of preconditioning. A rational explanation for these differences must await the results of more research.

Some of the effects of drought stress on growth of plants can be explained by the effects of ABA concentrations. A common response to elevated ABA concentrations is a reduction of shoot growth, which may be a result of reduced turgor rather than an increase in concentration of ABA. Changes in apical dominance correlated with water-stress-induced ABA levels have an effect on dormancy of axillary buds and terminal meristems. For example, with an increase in ABA concentrations in the terminal inflorescence of corn as a result of water stress, the number of axillary shoots produced increases. The stress related increase in ABA in the cereal grains leads to early ripening and the cessation of growth. There are also effects on flowering that are inhibited in some plants but initiated in others as a result of water stress and changes in ABA concentrations.

The effects of ABA related responses of roots to water stress are not clear. There is an effect on ion uptake because exogenous application of ABA inhibits movement of ions across the root but not absorption into the cortical cells. The effect on ion flux into the xylem from the pericycle is stimulatory because evidence indicates that xylem exudation of ions is increased. Studies of the resistances of movement of water across the root, and thus some ions as well, indicates that ABA decreases resistance and thus increases symplastic movement of water. The relationship of these responses to drought stress is not clearly defined yet.

INTERACTIONS OF PHYTOHORMONES IN DROUGHT STRESS

Three of the phytohormones play a major role in controlling plant water potentials during water stress. Water stress causes an initial increase in ETH concentration followed by a considerable increase in the concentration of ABA and a reduction in CK concentration. The increase in ABA concentration prevents further build up in ETH while CK, which can stimiulate ETH synthesis, declines in concentration which further exacerbates the lowering of ETH concentrations. One of the roles of ABA is modification of membrane permeability of the stomatal guard cells and higher concentrations resulting from water stress cause partial or complete closure of the stomates. The increased water potential that results has a moderating effect on any further changes in hormonal balances.

Prolonged water stress can result in leaf abscission. In such a situation the synthesis of ETH may be the predominate process and increases of ETH concentration cause a reduction in transport of auxin from the lamina to the abscission area of the petiole. Leaf abscission reduces leaf surface area and the plant is better able to reestablish a favorable water balance.

PHYTOHORMONE RESPONSE AND TEMPERATURE

As you may recall from reading Chapter 4, whether a given temperature is stressful to a plant or not depends upon the species, plant organs or structures involved, and the stage of development in the life cycle. Therefore, it is difficult to define specific temperature ranges that are considered stressful for plants in general but, suboptimal temperatures at or near freezing or supraoptimal temperatures above 35 to 40°C may be injurious to a number of plants, particularly herbaceous ones. Although such extremes in temperatures are generally considered to be harmful they also represent important environmental signals that may aid the plant in preparing for a stress as, for example, in the induction of dormancy. Such temperature signals may be necessary for the completion of a plant's life cycle as, for example, the cold induction of flowering (vernalization), vegetative growth, seed germination, and the breaking of dormancy. In some plants a heat treatment is required before flowering can be initiated. Plants such as tobacco and tomato grow best if subjected to differing day/night temperature regimes. If grown continuously at a single temperature they do not grow as readily. Thus, such plants may be experiencing a temperature stress. Since plant development is so closely linked to environmental temperature it is difficult to separate the stressful effects of temperature on plant hormone levels from those that may be operative in, and necessary for, growth, development, and acclimation processes. However, it is clear that plant hormone concentrations are sensitive to temperature and undergo changes in concentration and translocation pattern in response to temperature fluxes (Figures 10.2 and 10.3). Warm season crops such as bean, corn, eggplant, muskmelon, and okra exhibit elevated levels of free ABA and conjugated forms of ABA when grown at 10°C as compared to 25 or 40°C (Daie et al., 1981). In cool season crops such as beet, cabbage, lettuce, pea, and radish, only peas have higher ABA levels at 40°C than at 10 or 25°C. No difference in ABA concentration was observed for any of the other cool season crops growing at high or low temperatures. However, tomato seedlings exposed to day/night temperature regimes above optimum and below optimum contained increased concentrations of free and conjugated forms of ABA (Daie and Campbell, 1981). Because there was no change in water potential of the tomato plants at the different temparature regimes, water stress was not a factor in the increase of ABA concentrations.

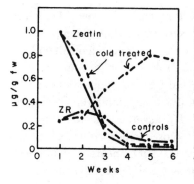

Figure 10.2. Effect of cold treatment on zeatin and zeatin riboside levels in wheat. Redrawn from F. Wightman, 1979. Modern chromatographic methods for the identification and quantification of plant growth regulators and their application to studies of the changes in hormonal substance in winter wheat during acclimation to cold stress conditions. In T. K. Scott (Ed.), *Plant Regulation and World Agriculture.* Used with permission from Plenum Publ. Corp., New York.

Exposure of organs of plants to temperature stress also changes the levels of phytohormones. Two-minute heat treatments of 46 to 74 °C to roots of *Nicotiana rustica* and *Phaseolus vulgaris* caused a reduction in the level of CK in xylem exudates but an increase in ABA levels (Itai et al., 1973). As a result, shoot and root growth were reduced. Corn roots subjected to temperatures of 8, 13, 18, 23, 28, and 33°C were found to export CK- and GA-like substances most readily at 28°C (Atkin et al., 1973). Inhibitor export was lowest at 33°C but increased at 3°C while GA and CK concentrations were at the lowest level at 3°C.

In their investigations of the effects of temperature on the phytohormone concentrations in cucumber fruits, Wang and Adams (1982) reported that a chilling temperature of 2.5°C accelerated the synthesis of ACC (the presursor from which ETH is synthesized) compared to that of fruits stored at 13°C. The skin tissue was found to have higher levels of

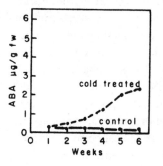

Figure 10.3. Effect of cold treatment on abscisic acid levels in wheat. Redrawn from F. Wightman, 1979. Modern chromatographic methods for the identification and quantification of plant growth regulators and their application to studies of the changes in hormonal substance in winter wheat during acclimation to cold stress conditions. In T. K. Scott (Ed.), *Plant Regulation and World Agriculture.* Used with permission from PLenum Publ. Corp., New York.

ACC and seemed to be more responsive than cortical tissue. Maximum ETH synthesis occurred 4 days after exposure to the chilling temperature and disappeared after 6 days. ACC levels were maximum 7 days after chilling and dropped to low levels after 9 days. Prolonged chilling appeared to damage the system that converts the ACC to ETH. Extremes in temperature seem to inhibit ETH synthesis primarily through inhibition of enzyme activity involved in its synthesis. Because ETH synthesis is often found to increase when plant tissues are chilled or exposed to heat stress and subsequently returned to unstressful temperatures, it appears that physical damage to cell and/or organelle membranes could decompartmentalize the enzymes of ETH synthesis causing an increase in its production.

PHYTOHORMONE RESPONSE AND NUTRITION

Deficiency and toxicity levels of the nutrient elements required by plants have been implicated in phytohormone regulation that leads to developmental alterations. The amount and source of nitrogen available for the sythesis of amino acids directly affects the synthesis of the phytohormones CK, ETH, and IAA, which are all derived from amino acids. Examples of the effect of nitrogen source and concentration on phytohormone concentrations are many. Sunflower plants grown with nitrate-N have higher levels of zeatin and zeatin ribosides than plants supplied with an ammonium source of nitrogen (Moorby and Besford, 1983). Birch seedlings supplied with low levels of ammonium nitrate had low levels of CK-like compounds in the leaves, while similar treatments to sycamore resulted in only a small reduction in CK levels (Darrall and Wareing, 1981). In the presence of ammonium sulfate the birch seedlings had little lateral shoot growth and tissues had no detectable CK activity compared to those grown with nitrate in the nutrient medium. The relationship between nitrate levels and the production of CK has been clearly demonstrated in potato plants in which nitrate-supplemented plants had increased amounts of CK in root exudates compared with plants grown in nitrate deficient media (Marschner, 1983). Because of the effect of molybdenum as a part of the nitrate reductase enzyme, the availability of this element effects the phytohormone concentrations through its effect on the snythesis of the amino acid precursors. Correlations between CK synthesis, flowering, and phosphorus levels have been observed in tomato, wheat, apple, and birch (Figure 10.4).

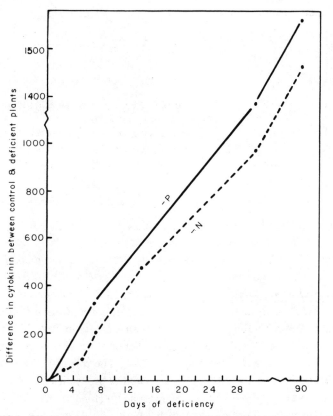

Figure 10.4. Effect of nitrogen and phosphorus deficiency on base and riboside cytokinin levels in birch leaves. Redrawn from J. M. Horgan, and P. F. Wareing, 1980. Cytokinins and growth responses of seedlings of *Betula pendula* Roth. and *Acer pseudoplatanus* L. to nitrogen and phosphorus deficiency. *J. Exp. Bot.* 31:525–532. Used with permission from Oxford Univ. Press, Oxford, UK.

Zinc is another element that is involved in the nitrogen nutrition of plants and phytohormone synthesis. Zn is required for the synthesis of tryptophan, which is a precursor of IAA. In tomatoes, for example, Zn deficiency symptoms can be eliminated either by addition of tryptophan or Zn. The little leaf disease of some fruit trees is caused by a Zn deficiency and the lack of IAA for cell enlargement of the leaves and the stem internodes. A symptom characteristic of this disorder is the formation of a rosette of

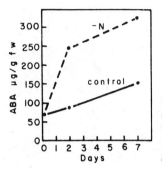

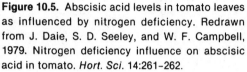

Figure 10.5. Abscisic acid levels in tomato leaves as influenced by nitrogen deficiency. Redrawn from J. Daie, S. D. Seeley, and W. F. Campbell, 1979. Nitrogen deficiency influence on abscisic acid in tomato. *Hort. Sci.* 14:261–262.

small leaves on the tree branches. The concentrations of ABA are also affected by nitrogen nutrition as shown in Figure 10.5.

There is much controversy over the role of boron in plant metabolism. A deficiency of B has been reported to depress CK concentrations but increase IAA production. Necrosis caused by B deficiency may be the result of IAA accumulation in tissues to a toxic concentration. The effect is caused by the reduction in IAA oxidase, an increase in bound IAA, and a reduction in free IAA. Boron has also been implicated in other aspects of metabolism, transport, and activity of IAA. The role of Ca in stabilizing membranes and their differential permeability properties has been known for many years. The role of Ca in membrane–phytohormone interactions is also important to the physiology of the plant. Ca deficiencies disrupt the integrity of membranes in such a manner that the synthesis or action of phytohormones at receptor sites on the membrane surface is altered. For example, IAA-induced hydrogen ion secretion, which is a prerequisite to IAA-induced growth in oat coleoptiles, is totally dependent on the presence of Ca in the medium. Ca has also been shown to stimulate ACC conversion to ETH and uptake of ACC in mung bean and cucumber hypocotyls (Ferguson, 1983). Frequent observations concerning IAA-induced ETH synthesis in a number of plant systems suggests a close association between Ca, IAA, ETH, membrane permeability, and subsequent control of growth of plant cells (Figure 10.6).

Toxic concentrations of nutrient elements can also affect phytohormone concentrations and action. High concentrations of copper and iron can increase ETH levels in some plants causing abscission of leaves. However, elevated amounts of cobalt inhibit ETH production in mung bean, cucum-

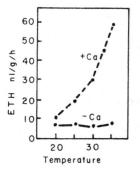

Figure 10.6. Effect of temperature and calcium on ethylene levels in mung bean hypocotyls. Redrawn from I. B. Ferguson, 1983. Calcium stimulation of ethylene production induced by 1-aminocyclopropane-1-carboxylic acid and indole-3-acetic acid. *J. Plant Growth Regul.* 2:205–214. Used with permission of Springer-Verlag, New York.

ber hypocotyls, and in apple fruit mesophyll. Other relationships of nutrient toxicities and phytohormone physiology no doubt exist but await further research to elucidate them.

Not only do deficient and toxic levels of ions of the nutrient elements affect phytohormone balances, but phytohormones such as IAA and CK can also direct the translocation patterns of the elements, an effect that shows the interdependence of phytohormones and nutrients in the regulation of plant growth and development.

PHYTOHORMONE RESPONSE AND PHOTOPERIOD

Photoperiod is a very important environmental trigger in plant morphogenesis. Like temperature, photoperiod controls many developmental aspects of plant growth required for the completion of the plant's life cycle. Although several light-sensing mechanisms may be operative in plants, the pigment, phytochrome, has been linked to several photoperiodic responses. Phytochrome is found in virtually all organs of higher plants. It is particularly prevalent in etiolated tissues but also occurs in photosynthetic systems and appears to be localized in membranes of etioplasts, mitochondria, and cytoplasm, and may be associated with cytoplasmic membranes. The phytochrome molecule consists of a protein and a chromophore that is sensitive to red and far-red wavelengths of light. In the presence of an abundance of red wavelengths of light (daylight, for instance), the red-absorbing form of the pigment (Pr) is converted to the far-red-absorbing form (Pfr) via a rearrangement of a double bond in the

TABLE 10.3 Phytochrome Mediated Plant Responses

Leaf expansion
Light inhibition or stimulation of seed germination
Flowering, photoperiodism
Light inhibition of root growth
Hypocotyl hook opening
Induction and breaking of dormancy of buds, tubers, and bulbs
Light inhibition of stem elongation
Light stimiulation of cereal leaf unrolling
Phototropisms
Geotropisms
Translocation

chromophore of the phytochrome molecule. Although Pfr can be converted back to Pr by exposure to far-red light, it is not completely clear how this conversion occurs. The dark period to which the plant is subjected is important in the degradation of Pfr and its conversion to Pr; there may also be a simultaneous degradation of Pfr and resynthesis of Pr in the dark. Pfr is the biologically active form of phytochrome. The photoperiodic regulation of the proportions of the two forms of phytochrome and their interaction with other environment-sensing systems in the plant that detect changes in the environment may induce enzyinatic synthesis of phytohormones responsible for controlling the processes of photomorphogenesis. Considerable gaps exist in our understanding of the link between the light initiated phytochrome reactions, the initiation of phytohormone reactions, and the control of specific growth responses.

Even though endogenous phytohormone concentrations respond to light, it is not clear that such changes are directly or indirectly controlled by phytochrome. Changes in phytohormone levels as a result of changing day lengths have been reported many times. Generally, long days promote synthesis of growth stimulators and short days promote synthesis of growth inhibitors. Table 10.3 lists several plant responses thought to be controlled by light-mediated phytochrome action and changes in the levels and/or ratios of phytohormones. All the classical growth horomones are known to respond to changing photoperiod; however, much of the information available focuses on IAA and GA, with less known about the effect of light on ETH, CK, and ABA.

Coleoptile, hypocotyl, and epicotyl segments of grasses and legumes have been used extensively to study light effects on IAA. Since excision of segments may remove them from important sources of phytohormones, care must be taken in interpreting such information. IAA decreases in tissues exposed to white light or red light (R). Increases in IAA concentrations occur in the dark or after exposure to far-red light (FR). Several instances of R inhibition of IAA oxidase have been reported that contradict observations regarding R effects on IAA levels. However, some phenolic substances, which increase in concentration in plants as a result of R treatment, are stimulatory while others inhibit IAA oxidase activity. Thus, a complex metabolic system involving interactions among phenolics, IAA oxidase, and IAA may occur in response to exposure to R.

Intact plants have also been studied with respect to the effects of light on IAA levels. Dark-grown bean plants contained higher levels of IAA and were taller compared to R treated plants, which were short and had low levels of IAA. Plants fed with [^{14}C]IAA resulted in the formation of three labeled fractions that were proportionally related to light quality.

Other light responses, such as initiation of hypocotyl hook opening, have been correlated with reduced levels of IAA. In addition, R reduces polar auxin transport in rice coleoptiles but enhances acropetal transport of IAA through root segments. Also, light (420 to 470 nm, blue to near UV) is effective in the lateral movement of IAA in avena coleoptiles causing a positive phototropism.

GA has been implicated with such light-mediated growth responses as leaf expansion, seed germination, unrolling of cereal leaves, and stem growth. Again, some of these responses seem to be under the control of phytochrome. It is not completely clear how phytochrome and GA interact to control many of these responses but the evidence indicates a direct correlation between phytochrome (Pfr) and the production of GA in leaves exposed to R and the expansion and unrolling of cereal leaves. The relationship appears to be different in stems. Many dwarf varieties of plants grow as well in the dark as their tall counterparts. However, exposure to light causes a dramatic reduction in growth compared to tall plants. The inhibition can be countered by the application of GA to the stems. Such a response suggests that light inhibits GA production and thus inhibits stem growth. However, analysis of these plants shows no difference in GA levels when grown in the dark or the light (Table 10.4). Differences in kinds and effectiveness of GA in these plants have been observed when grown under light or dark conditions. Since light-grown bean and pea plants have been observed to contain more ABA than dark-grown plants, light could sup-

TABLE 10.4 Extractable and Diffusable Levels of Gibberellic Acid from *Pisum sativum* Apical Buds

	Light		Dark	
	Normal	Dwarf	Normal	Dwarf
		(mg GA$_3$ equiv./kg FW)		
Diffusion				
GA$_1$	1.7	2.30	1.90	2.30
GA$_5$	0	0	0	0
Total	1.7	2.30	1.90	2.30
Extraction				
GA$_1$	0.90	0.60	1.00	0.80
GA$_5$	0.40	0.45	0.52	0.36
Total	1.30	1.05	1.52	1.16

Source. M. Black and A. J. Vlitos. 1972. Possible interrelationships of phytochrome and plant hormones. In K. Mitrakos and W. Shropshire, Jr. (Eds.), *Phytochrome.* Used with permission from Academic Press, New York.

press the sensitivity of stem tissues to differing GA, cause the synthesis of effective GA, or cause the formation of growth inhibitors.

The relationships of light, phytochrome, and GA in seed germination are less clear and seemingly more complex. In many seeds, light is required after imbibition for germination to occur. If a light treatment is followed by FR treatment, germination is nullified. In many cases GA can substitute for the light treatment and induce germination but not all light requiring seeds respond to GA treatment. However, even in those seeds that do not respond to GA treatment, germination is still phytochrome mediated, which suggests that other pathways of action are operative or that the synthesis of a specific GA is required.

The evidence for R-phytochrome-mediated production of ETH and ABA is even less clear. There is some evidence that hypocotyl hook opening is controlled by light inhibition of ETH production, which is FR reversible. Also, exogenous ETH enhances R treated lettuce seed germination when the seeds are kept in the dark, but there is no clear evidence that ETH production is directly mediated through changes in phytochrome response to R. However, increase in ETH production may be a secondary response reflecting changes in metabolic activity initiated through phytochrome activity. Whether directly or indirectly mediated through phytochrome, ETH production does respond to light (Figure 10.7).

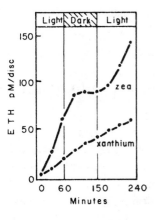

Figure 10.7. Effect of light on ethylene production in leaf discs of corn and cocklebur. Redrawn from B. Grodzinski, I. Boesel, and R. F. Horton, 1982. The effect of light intensity on the release of ethylene from leaves. *J. Exp. Bot.* 33:1185–1193. Used with permission from Oxford Univ. Press, Oxford, UK.

Few responses indicate ABA production is under phytochrome control. Light is known to inhibit root growth with a subsequent increase in ABA and, in addition, light-grown bean and pea plants have higher levels of ABA than dark-grown plants. The difference in ABA levels may be related more to changes in plant water potentials than to the presence or absence of light. However, in pearl millet, light was stimulatory to ABA production and, in some instances, allowed an accumulation of ABA independently of the rate of photosynthesis (Henson, 1983).

Evidence for a role of CK in phytochrome-related responses is scarce. CK substitutes for R in wheat leaf unrolling, in expansion of etiolated leaves of some plants, and in growth of the duckweed, Lemna. The level of CK fluctuates diurnally in poplar leaves with peak activity occurring at daybreak and a reduced activity at night. However, in rose buds kept in the dark by wrapping them in aluminum foil, the levels of CK increased in the upper shoot tips. When the foil was removed and the buds were exposed to light for 2 days, CK concentrations fell.

It is clear from the foregoing discussion that light does have an effect on phytohormone balances, but it is not clear what the biochemical mechanisms are for control of phytohormone levels. Evidence for phytochrome-mediated control of IAA and GA reactions is stronger than that for ETH, CK, or ABA reactions. Thus, when light is limiting or in excess and when normal photoperiods and light quality are altered, one is likely to observe stress responses that are manifested through a multitude of changes in physiological processes. Chapter 7 contains additional information on the effects of irradiation stress on plants.

PHYTOHORMONE RESPONSE TO PATHOGENS AND INSECTS

Bacterial, fungal, and viral infections of plant tissues often result in elevated levels of endogenous phytohormones. Many of the symptoms associated with infected plants are hormone mediated and, in some instances, can be duplicated through phytohormone manipulation. Some of the symptoms of infection include epinasty, adventitious root formation, chlorosis, xylem hyperplasia, tylosis formation, shoot elongation, wilting, and stunting. Plant pathogenic organisms are known to produce phytohormones *in vivo* as well as cause changes in phytohormone balance as a result of the infection process. Changes in phytohormone balances in infected plants are usually through the alteration of the host's synthetic processes and may be caused by substances produced by the pathogen that directly alter physiological and biochemical processes in the plant.

The symptom in which a green island of leaf tissue is surrounded by a chlorotic ring of tissue can be induced in a number of plants by obligate parasites that cause the diseases of mildew, downy mildew, and rust. The formation of green islands could represent the induction of CK synthesis in these areas and the consequent retention or resynthesis of chlorophyll as well as the mobilization of nutrients to the infection site. Such a role for CK has been demonstrated with applications of CK that delay senescence in detached leaves and also increase translocation of nutrients to the sites of application. Green islands have been observed also as a symptom of infestation with parasitic insects such as leaf minors. Analysis of chlorotic zones of infected tissues surrounding the infection zones indicate significant levels of CK are present. Because other hormones are involved in leaf senescence the interactions may be considerably more complicated than indicated.

Diseases resulting in elevated levels of CK activity in plant tissues are included in Table 10.5.

Plant tissues infected with bacteria, fungi, and viruses generally exhibit increased levels of IAA. A few reports of viral infections of tomato, bean, and beets (curly top virus) indicate IAA levels are reduced. At present the

TABLE 10.5   Plant Diseases Characterized by Elevated Cytokinin Levels

Disease	Incitant	Symptoms
Fasciation disease	*Corynebacterium fascians*	Increase in numbers of shoots, shoots becoming thickened, fleshy and misshapen.
False broom rape	*Corynebacterium fascians*	Roots characterized by large white fleshy outgrowth from topped and debudded plants; occurs in tobacco roots.
Crown gall	*Agrobacterium tumefaciens*	Tumor formation in a large number of species and families, mostly dicotyledenous.
Clubroot	*Plasmodiophora brassicae*	Roots and hypocotyl cells undergo abnormal enlargement and dense root gall formation with subsequent negative geotropism; occurs in cruciferae roots.
Rust gall	*Cronartium fusiforme*	Gall formation of pine stems.
Root knot	Meloidogyne sp.	Root disease caused by several species of Meloidogyne. Stunted growth, root gall formation, wilting, giant cell formation.

role of bacteria, fungi, and viruses in the induction of IAA in plants is unclear, but some of the alterations and changes that occur include:

1. Increased synthesis of IAA precursors.

2. Stimulation of host IAA oxidases and phenolic substances that can either inhibit or stimulate IAA oxidases.

3. Secretion of IAA oxidases from the pathogen.

4. Inhibition of auxin-degrading enzymes through the production of phenols, phenol oxidases, and hydroxylases as a result of infection.

5. Creation of a water stress, which may alter phytohormone balances.

Interactions of IAA with ETH should also be considered when physical damage of cells occurs as a result of infection.

ETH production is stimulated by infection primarily as a result of tissue injury. Symptoms of increased ETH production include increased membrane permeability, epinasty, inhibition of root growth, induction of senescence, leaf and fruit abscission, and increased diameter growth of stems. ETH produced as a result of wounding upon infection can also initiate the production of specific peroxidases, polysaccharide hydrolases, chitinase, and polyphenoloxidases, which may contribute to degenerative processes in cells. In a few instances, however, ETH may enhance disease resistance by causing the production of antifungal quinones derived from the oxidation of phenolic acid compounds. Infection may also lead to the production of hydrogen peroxide, which is damaging to plant tissues. However, as ETH increases, more peroxidases are induced which destroy hydrogen peroxide.

One of the most documented examples of the involvement of GA in disease response can be attributed to the discovery that a fungal disease of rice, the "bakanae" disease caused by *Gibberella fujikuroi* brings about severe elongation of stems followed by lodging. Initially it was thought that a substance, later identified as GA, was produced by the fungus and was responsible for the elongation growth. However, the organism also stimulated the host to increase its own endogenous levels of GA. Both processes contribute to an increase in elongation growth and the discovery subsequently led to the research that showed that GA is a phytohormone of widespread distribution and activity. Both fungi and bacteria are known to produce GA-like substances *in vivo*, however, more is known about the fungal effects on GA production in plants than the bacterial effects. Diseases causing stunting, such as mosaic virus infections of cucumber and tobacco, have been correlated with reduced levels of GA in infected tissues.

The accumulation of inhibitors in diseased plant tissues has been associated with stunting of growth. Although production of the inhibitor, ABA, has been observed, it is not clear whether its production is in direct response to infection or in response to secondary effects, such as water stress, that often accompany infection.

The feeding of insects and mites on plant tissues also cause changes in endogenous levels of phytohormones as well as contribute phytohormones from their saliva. A form of abberant growth associated with insects is the formation of galls of various shapes and sizes. The growth has been correlated with increased production of IAA and CK or, as with the caterpillar of *Stigmella argentipedella*, secretion of substances such as CK in saliva. The mobilization of nutrients to the feeding area helps the caterpillar in its growth and metabolism. Many insects cause abscission of plant parts by causing changes in the phytohormone balance and the wounding by the feeding insect causes production of ETH, which is involved in the abscission process.

DISCUSSION QUESTIONS

1. What are the principle phytohormones that play a role in plant responses to environmental stress?

2. IAA concentration in plant tissues decreases because of water stress. What are some possible explanations?

3. In what aspect of water stress is GA important?

4. Someone has said that it is easy to cause a plant to produce ethylene—just kick it! How can you relate this statement to production of ethylene by plants under water stress?

5. What is the series of events related to water stress that result in abscision of leaves?

6. What is the correlation between concentration of ABA and CK during water stress? What role do ABA and CK play in control of drought stress tolerance?

7. As a result of the increases in ABA concentration during drought stress, what are some morphogenetic changes that occur that make the plant more tolerant?

8. Indicate four ways in which the internal levels of ABA may be controlled during water stress.

9. Compare warm season crops to cool season crops with respect to ABA concentration when grown under conditions of temperature stress.

10. By what means do temperature extremes reduce ETH concentrations? On the other hand, chilling temperatures and high temperatures cause an increase in ETH. What are some possible explanations?

11. Why is it sometimes difficult to define a temperature range considered to be stressful to plants?

12. There are interactions between nitrogen nutrition and stress levels of phytohormones. Explain. What is the role of zinc in phytohormone synthesis?

13. Some of the inorganic nutrients have direct effects on membrane permeability. Which ones are they and how do they change permeability?

14. Deficient and toxic levels of ions can directly affect the synthesis of phytohormones. Conversely, in what manner do phytohormones affect ions?

15. Name some light mediated growth responses attributable to IAA and GA.

16. Would you consider light inhibition of stem elongation a stress response? Explain.

17. What effect does light have on phytohormone control of stress responses? Give some examples.

18. Visual symptoms of disease are often similar to the effects of phytohormones. List some symptoms and relate them to changes in phytohormone levels that occur.

19. Which phytohormone was discovered as a result of studies of diseased rice plants?

REFERENCES

Addicott, F. T. 1982. *Abscission*. University of California Press, Berkeley, pp. 185–216.

Aspinall, D. 1980. Role of abscisic acid and other hormones in adaptation to water stress. In N. C. Turner and P. J. Kramer (Eds.), *Adaptation of Plants to Water and High Temperature Stress*. Wiley-Interscience, New York, pp. 155–172.

Atkin, R. K., G. E. Barton, and D. K. Robinson. 1973. Effect of root-growing temperature on growth substance in xylem exudate of *Zea mays*. *J. Exp. Bot.* 24:475–487.

Black, M., and A. J. Vlitos. 1972. Possible interrelationships of phytochrome and plant hormones. In K. Mitrakos and W. Shropshire, Jr. (Eds.), *Phytochrome*. Academic, New York, pp. 517–550.

Bradford, K. J., and T. C. Hsiao. 1982. Physiological responses to moderate water stress. In O. L. Lange, P. S. Nobel, C. B. Osmond, and H. Ziegler (Eds.), *Physiological Plant Ecology*, Vol. 2. Springer-Verlag, New York, pp. 264-324.

Chanan, I., A. Ben-Zioni, and L. Ordin. 1973. Correlative changes in endogenous hormone levels and shoot growth induced by short heat treatments to the root. *Physiol. Plant.* 29:355-360.

Crawford, R. M. M. 1982. Physiological responses to flooding. In O. L. Lange, P. S. Nobel, C. B. Osmond, and H. Ziegler (Eds.), *Physiological Plant Ecology*. Vol. 2. Springer-Verlag, New York, pp. 453-477.

Daie, J., and W. F. Campbell. 1981. Response of tomato plants to stressful temperatures. *Plant Physiol.* 67:26-29.

Daie, J., S. D. Seeley, and W. F. Campbell. 1979. Nitrogen deficiency influence on abscisic acid in tomato. *Hort. Sci.* 14:261-262.

Daie, J., W. F. Campbell, and S. D. Seeley. 1981. Temperature stress-induced production of abscisic acid and dihydrophaseic acid in warm- and cool-season crops. *J. Am. Soc. Hort. Sci.* 106:11-13.

Darrall, N. M., and P. F. Wareing. 1981. The effect of nitrogen nutrition on cytokinin activity and free amino acids in *Betula pendula* Roth. and *Acer pseudoplatanus* L. *J. Exp. Bot.* 32:369-379.

DeGruf, J. A., and H. Frederick. 1983. Photomorphogenesis and hormones. In W. Shropshire, Jr. and H. Mohr (Eds.), *Photomorphogenesis*, Vol. 16A. Springer-Verlag, New York, pp. 401-427.

Dekhuijzen, H. M. 1976. Endogenous cytokinins in healthy and diseased plants. In R. Heitefuss and P. H. Williams (Eds.), *Physiological Plant Pathology*. Springer-Verlag, New York, pp. 256-550.

Dugger, W. M. 1983. Boron in plant metabolism. In A. Lauchli and R. L. Bieleski (Eds.), *Inorganic Plant Nutrition*, Vol. 15B. Springer-Verlag, New York, pp. 626-650.

Ferguson, I. B. 1983. Calcium stimulation of ethylene production induced by 1-aminocyclopropane-1-carboxylic acid and indole-3-acetic acid. *J. Plant Growth Regul.* 2:205-214.

Field, R. J. 1981. The effect of low temperature on ethylene production by leaf tissue of *Phaseolus vulgaris* L. *Am. J. Bot.* 47:215-223.

Grodzinski, B., I. Boesel, and R. F. Horton. 1982. The effect of light intensity on the release of ethylene from leaves. *J. Exp. Bot.* 33:1185-1193.

Henson, I. E. 1983. Effects of light on water stress induced accumulation of abscisic acid in leaves and seedling shoots of pearl millet (*Pennisetum americanum* Luke). *Z. Pflanzenphysiol.* 112:257-268.

Horgan, J. M., and P. F. Wareing. 1980. Cytokinins and the growth responses of seedlings of *Betula pendula* Roth. and *Acer pseudoplatanus* L. to nitrogen and phosphorus deficiency. *J. Exp. Bot.* 31:525-532.

Itai, C., and Y. Vaadia. 1971. Cytokinin activity in water stressed shoots. *Plant Physiol.* 47:87-90.

Itai, C., A. Ben-Zioni, and L. Ordin. 1973. Correlative changes in endogenous hormone levels and shoot growth induced by short heat treatments to the root. *Physiol. Plant.* 29:355-360.

Kramer, P. J. 1980. Drought stress and the origin of adaptations. In N. C. Turner and P. J. Kramer (Eds.), *Adaptation of Plants to Water and High Temperature Stress*. Wiley-Interscience, New York, pp. 7-20.

Levine, A. 1972. Water deficits and hormone relations. In T. T. Kozlowski (Ed.), *Water Deficits and Plant Growth*, Vol. 3. Academic, New York, pp. 255-275.

Levitt, J. 1980. *Responses of Plants to Environmental Stresses*. Vol. 2. Academic, New York, pp. 25-92.

Marme, D. 1983. Calcium transport and function. In A. Lauchli and R. L. Bieleski (Eds.), *Inorganic Plant Nutrition*, Vol 15B. Springer-Verlag, New York, pp. 599-625.

Marschner, H. 1983. General introduction to the mineral nutrition of plants. In A. Lauchli and R. L. Bieleski (Eds.), *Inorganic Plant Nutrition*, Vol. 15. Springer-Verlag, New York, pp. 5-60.

Mengel, K., and E. A. Kirkley. 1982. *Principles of Plant Nutrition*, 3rd ed. International Potash Inst., Atlanta, pp. 501-511, 533-543.

Milborrow, B. V. 1981. Abscisic acid and other hormones. In L. G. Paleg and D. Aspinall (Eds.), *The Physiology and Biochemistry of Drought Resistance in Plants*. Academic, New York, pp. 347-388.

Moorby, J., and R. T. Besford. 1983. Mineral nutrition and growth. In A. Lauchli and R. L. Bieleski (Eds.), *Inorganic Plant Nutrition*, Vol. 15B. Springer-Verlag, New York, pp. 481-527.

Pegg, G. F. 1976. Endogenous auxins in healthy and diseased plants. In R. Heitfuss and P. H. Williams (Eds.), *Physiological Plant Pathology*. Springer-Verlag, New York, pp. 560-581.

Pegg, G. F. 1976. The involvement of ethylene in plant pathogenesis. In R. Heitefuss and P. H. Williams (Eds.), *Physiological Plant Pathology*. Springer-Verlag, New York, pp. 582-591.

Pegg, G. F. 1976. Endogenous gibberellins in healthy and diseased plants. In R. Heitefuss and P. H. Williams (Eds.), *Physiological Plant Pathology*. Springer-Verlag, New York, pp. 592-606.

Pegg, G. F. 1976. Endogenous inhibitors in healthy and diseased plants. In R. Heitefuss and P. H. Williams (Eds.), *Physiological Plant Pathology*. Springer-Verlag, New York, pp. 607–616.

Smith, H. 1975. *Phytochrome and Photomorphogenesis*. McGraw-Hill, New York, pp. 139–159.

VanStaden, J., N. Zieslin, H. Spiegelstein, and A. H. Halevy. 1981. The effect of light on the cytokinin content of developing rose shoots. *Ann. Bot.* 47:155–157.

Wang, C. W., and D. O. Adams. 1982. Chilling induced ethylene production in cucumbers (*Cucumis sativus* L.). *Plant Physiol.* 69:424–427.

Wightman, F. 1979. Modern chromatographic methods for the identification and quantification of plant growth regulators and their application to studies of the changes in hormonal substance in winter wheat during acclimation to cold stress conditions. In T. K. Scott (Ed.), *Plant Regulation and World Agriculture*. Plenum, New York, pp. 327–377.

Wright, S. T. C. 1978. Phytohormones and stress phenomena. In D. S. Letham, P. B. Goodwin, and T. J. V. Higgins (Eds.), *Phytohormones and Related Compounds—A Comprehensive Treatise*, Vol. 2, Elsevier/North-Holland, Amsterdam, pp. 495–532.

11

STRESS TOLERANCE THROUGH BIOTECHNOLOGY

The application of chemicals such as plant growth regulators (PGR) may alter the expression of the genome of plants in such a way that they are tolerant of specific stresses. Such changes are not heritable. The use of the technology of genetic engineering can change the genome itself and produce a heritable trait that may render a plant tolerant of a specific stress. These are but two of the technologies available, but are the ones that have been investigated to the greatest degree in reference to stress tolerance.

USE OF PLANT GROWTH REGULATORS

The use of chemicals to alleviate stress and avoid injury or death has been investigated only on a limited scale, but interest is increasing as possibilities for success increase in direct relation to the increased understanding of the physiological affects of stress. Alteration of growth and development with plant growth regulators (PGR) so that resistance is increased is possible and based on sound physiological principles. Some examples of the type of research that is being done are recounted in the the discussion that follows.

Increasing Drought Tolerance

One of the ways that drought stress can be reduced is by decreasing the rate of transpiration or increasing the rate of absorption of water by inducing root growth or greater absorption efficiency. Little research has been done on the latter aspect of controlling water use efficiency but research concerning more direct means of controlling transpiration has been carried out for several years. The principal methods that have been investigated are:

1. Use of antitranspirant chemicals that form a film over the surface of the leaf and effectively block transpiration.

2. Use of chemicals that cause stomates to partially or wholly close.

3. Use of growth retardants or inhibitors that change the morphology of the plant by reducing leaf size or number, change root to shoot ratios, reduce the number of stomates per unit area of leaf, or otherwise alter growth patterns in favor of greater water use efficiency.

4. Increase of the reflectance of irradiation by applying a reflecting substance to the foliage.

Film-forming antitranspirants are usually polymers such as polyvinyl waxes or polyethylene and vinylacrylate. Alcohols such as hexadecanol have been used also. These types of antitranspirants function by physically preventing water vapor from leaving the internal tissues of the leaf, but they also interfere with the exchange of CO_2 and O_2. The result is a change in photosynthesis and other metabolic processes in the leaves. Transpiration rates have been successfully reduced by 30 to 50% with an average coating of 50% of the leaf surface with no measurable impairment of photosynthesis.

Control of gas exchange can be achieved by using metabolic antitranspirants at various concentrations. However, a limitation of these types of antitranspirants is that their activity extends beyond guard cells to mesophyll cells. The impairment of photosynthesis which may result is detrimental to the plant. Some of the most promising antitranspirants of the metabolic type are similar to ABA or have anti-CK activity. ABA is not effective as a commercial antitranspirant because it is rapidly metabolized, is unstable in UV light, and is expensive to synthesize in quantity. Natural metabolites and compounds resembling ABA (i.e., phaseic acid, dihydrophaseic acid, xanthoxin, and vomifoliol) also have limited usefulness. Anticytokinins such as 4-cyclopentylamino-2-methylthio-pyrolo(2,3-*d*)-pyrimidine and 3-methyl-7-pentylaminopyrazoli(4,3-*d*)-pyrimidine are antagonists to cytokinin activity and prevent stomatal opening. ABA and anticytokinins have been shown to act synergistically to reduce transpiration when combined and applied to detached wheat leaves, but foliar sprays applied to intact plants were not effective. Other chemicals with less specific activity on stomatal movement include certain fatty acids, phenylmercuric acetate, alkenylsuccinic acid, salicylaldoxime, and some inhibitors of photosynthesis.

The use of chemicals to modify morphology by selectively changing growth rates of plant organs has been another way of achieving reduced transpiration rates. Chemicals such as chlorocholine chloride (CCC), AMO-1618, (ammonium (5-hydroxycarvacryl)trimethylchloride piperidine hydrochloride), CBBP (2,4 dichlorobenzyltributylphosphonium chloride), and SADH (succinic acid 2,2-dimethylhydrazide) have met with varied success. All inhibit height growth (possibly by inhibiting GA), and are frequently reported to reduce leaf area, reduce the density of stomates, increase the thickness of leaves, and produce a greater root to shoot ratio. Such modifications of growth could lead to reductions in transpiration rates, but often lead to reduced production.

A role in changing membrane permeability and absorption rate for water is another possible mechanism of water conservation in plants that remains to be explored more thoroughly. For example, CCC is a choline derivative and choline is an important component of phospholipids, which comprise plant membranes. Some phospholipids may act as ion carriers across membranes and, thus, treatment with CCC may affect the formation or functioning of phospholipids, which act as ion carriers. Table 11.1 lists some of the effects of growth retardants on controlling water stress in a variety of species.

Cold Tolerance

The most frequently used chemicals in attempts to increase tolerance to low temperatures have been SADH and CCC, but there has been little research since the 1960s. There has been some success in improving cold tolerance of such plants as cabbage, tomato, box elder seedlings, young

TABLE 11.1 Drought Tolerance Induction with Growth Retardant

Plant	Response	Growth Retardant
Wheat	More shoots produced after rewatering	SADH & CCC
Barley	Used ½ H_2O as controls/more grain	CCC
Grape	Reduced water usage, less wilting, greater tolerance of drought	SADH
Bean	Reduced wilting	Phosfon & CCC
Apple	Water deficits lowered in fruits and leaves	SADH
Sunflower	Improved ability to recover from drought, improved yields	SADH & CCC
Gladiolus	Increased dry matter of corms under water stress	CCC & AMO-1618
Corn	Decreased water usage, reduced transpiration in drought susceptible but not drought tolerant corn	CCC

pear trees, wheat, raspberry, strawberry, and azalea. Applications of very high concentrations (1000 to 10,000 ppm) were used as foliar sprays or as soil drenches. Seed treatment of wheat with CCC to improve cold tolerance of seedlings is a common practice in the USSR. Maleic hydrazide has been used to improve frost tolerance in citrus trees because it reduces defoliation and deblossoming. It has also been used to increase cold tolerance of red raspberry, grape, and mulberry. However, maleic hydrazide often causes leaf malformation as well, which has made the use of the chemical less desirable.

Chilling temperatures may produce water stress accompanied by an increase in ABA and a decrease in GA. Exogenous application of ABA to cotton seedlings and cotyledonary discs prior to chilling prevented injury when exposure to 4°C was less than 5 days (Rikin, 1979). ABA also increased cold tolerance of *Acer negundo* and Rome Beauty apple trees. GA synthesis is apparently associated with decreased cold hardiness and can be countered by application of ABA and other inhibitors.

The chilling resistance of *Pisum sativum* and *Raphanus sativus* has been reported to be increased by applications of benzyladenine (BA), a synthetic CK, possibly by increasing mobilization of hardening promoters such as ABA-like inhibitors or complex sugars. In general, CK increases tolerance to low- and high-temperature extremes. Polyamines, such as long-chained alkylene diamines, have been effective in decreasing cold or frost damage in such plants as wheat, rice, barley, oats, corn, millet, soybean, lima beans, pole beans, snap beans, peanuts, spinach, lettuce, tomato, mulberry, tobacco, fruit trees, poinsettia, carnation, and geranium.

Other chemicals with reported cold tolerance inducing characteristics are 2-amino-6 methyl benzoic acid (wheat, cereals, tobacco, and grapes), naphthalene acetic acid (orange trees), several analogs of 5-chloro-4-quinoline carboxylic acid (zucchini), mepiquat chloride (citrus), and polyethylene glycol.

Salt Tolerance

Increasing tolerance of plants to salt stress is of particular importance in dry land areas where crop irrigation is a necessity. The continued practice of irrigation almost always leads to increased salinity of soils in the rooting zone of crop plants. The development of PGR chemicals that are effective in increasing salt stress tolerance has a dual purpose: (1) to decrease transpiration rates with a subsequent reduction in irrigation needs and the salinity problem, and (2) to increase plant tolerance to saline soils with a subsequent expansion to land area that cannot now be cultivated.

Because salts in the soil lower the water potential and make water less available to plants, responses of plants to salt stress are frequently similar to the responses to drought stress. However, differences in the physiology of plants exposed to salt stress do occur. For example, in both types of stresses ABA concentration increases which causes stomates to close, but the CK levels decline. The ABA levels remain elevated for longer periods under salt stress than under drought stress even after the plants regain turgor. There may be different controlling systems responding to the two different stresses. If the systems are different, the use of PGR to control the plant reaction would also be different. To date the difficulty in separating the effects experimentally probably accounts for the sparsity of information on the use of PGR to alleviate the effects of salt stress. However, there are reports of the use of AMO-1618, phosfon, and CCC to increase salt tolerance in plants such as wheat, spinach, and soybean. In some instances, seed treatments with these compounds have been effective. More recently, naturally produced substances such as choline and betaine have been shown to act as "osmo-protectants" when *Escherichia coli* was subjected to osmotic stress (LeRudulier et al., 1984). The hardening of plants to drought stress or the use of PGR to increase tolerance to drought stress in some plants leads to tolerance of low-temperature stress or high-salt stress. Consequently, PGR used to control drought stress also may have application in alleviating salt stress.

Nutrient Stress

PGR can be used in several ways to improve nutrient uptake and the efficiency of use. (1) The physiology of the plant can be altered so that the rate and selectivity of uptake can be controlled. (2) PGR can be used to improve root growth and thus increase the volume of soil penetrated by roots. (3) Mycorrhizal associations with roots can be manipulated with a consequent effect on phosphorus and nitrogen nutrition. Also, chemicals are now available to specifically inhibit nitrifying bacteria that inhabit the soil so that the available nitrogen is increased because the rate of leaching is reduced.

The information that is available on the interaction of PGR with plant nutrition primarily addresses the effects on accumulation of nutrients in tissues following treatment with PGR. There is little known about what happens physiologically in the interactions. For example, GA has been reported to accelerate the uptake of K in wheat and 2,4-D stimulates the accumulation of N and P in wheat and soybean. However, both chemicals interfered with translocation of all three of the nutrients. SADH did not affect uptake, but stimulated translocation of N, P, and K. Treatment with CCC resulted in higher concentrations of N, Ca, and Mg in the stem tissues of tomato and SADH caused an increase in N only. In wheat, CCC treatment caused a reduction of nitrogen levels and a retardation of protein accumulation. When CCC was combined with 2,4-D, both N and protein accumulated and foliar application of CCC and SADH resulted in lower levels of N and P in the fruit of okra.

The effects of PGR on the uptake, translocation, and partitioning of nutrients under stress conditions need further investigation before such chemicals can be used profitably to increase production. As researchers continue to find ways to use marginal lands for crop production, the use of PGR must be seriously considered as a way of altering the physiology of crops that may have more immediate results than long-term breeding or genetic manipulation programs.

Air Pollutants

The information concerning the use of PGR to alleviate air pollution injury or to increase tolerance of crop plants to air pollutants is quite limited. Ozone exposure causes leaf injury, reduction of growth, and decreased yields of several crops such as potatoes, white beans, pinto beans, and tobacco. The chemical N-[2-(-oxo-1-imidazolidinyl), ethyl]-N'-phenylurea (EDU) has been reported to increase ozone resistance of pinto beans by 30-fold when applied as a foliar spray (Carnahan et al., 1978). Antisenescence chemicals such as benzimadazole, N-6-benzyladenine, and kinetin have reduced ozone damage in pinto bean and spinach. The increase in resistance was correlated with reduced losses of free sterols associated with the chloroplasts of the leaf tissue (Tomlinson and Rich, 1973).

USE OF GENETIC ENGINEERING

Recent advances in biotechnology used to explore and manipulate the genome of plants hold promise for increasing our understanding of mechanisms regulating stress responses. One method utilizes the bacterium *Agrobacterium tumefaciens*, which causes crown gall on many dicotyledenous plants. It is indeed ironic that a disease producing organism is being used as a primary tool to understand both biotic and abiotic stress inducing mechanisms.

The infection of a plant with agrobacterium results in the formation of gall-like swellings at the transition zone between the root and shoot. When the bacterium infects the tissues, circular segments of DNA called plasmids which encompass several genes are inserted into the host cell DNA. The bacterium alters the genome of the plant and reprograms its metabolism. The result is the sysnthesis of nutrients used by the bacterium for growth and the synthesis of hormones used by the plant causing tumor development at the site of the infection.

It is now possible by using the techniques of genetic engineering to remove plasmids from agrobacterium, modify them by inserting new DNA obtained from another organism, and transfer the inserted code for desirable traits to a target plant. To accomplish the transfer the new plasmid is inserted into an agrobacterium cell which is cultured in a manner that produces thousands of cells containing the new genetic information. A host plant is inoculated with the altered bacteria and the new genetic information is incorporated into the genome of the host plant where it may cause beneficial effects in terms of stress tolerance.

Other bacteria also are known to produce plasmids and have been used in a similar manner in recombinant DNA research. To improve water stress tolerance in some legumes, genes responsible for proline synthesis have been introduced into rhizobia species that have been allowed to colonize and form nodules on the roots of plants. The gene modification prevents the feedback inhibition of bacterial proline synthesis and high concentrations of proline are produced. In this relationship the genetics of the host plant are not altered but when the plant is subjected to drought stress the thousands of rhizobia associated with the root nodules are induced to synthesize proline that is available to the plant as an osmoprotectant.

Genetic engineering also has been used successfully to develop herbicide-resistant plants. The herbicide sold under the trade name "Roundup" is nonselective and may kill any plant on contact. To prevent injury or death of crop plants would render the herbicide more useful in weeding crops. The active ingredient, glyphosate, blocks the enzyme that causes the synthesis of essential aromatic amino acids. If the synthesis of the essential aromatic amino acids could be made to occur in the presence of glyphosate, plants would be rendered resistant to the herbicide.

The bacterium *Salmonella typhimurium* possesses the identical pathway for the synthesis of aromatic amino acids that occurs in higher plants. Through mutagenizing techniques a glyphosate-resistant mutant has been developed. The gene for resistance has been identified and has been incorporated into the Ti plasmid of *Agrobacterium tumefaciens*. Cultures of the bacteria containing the altered plasmid were used to inoculate leaf discs of tobacco, tomato, and poplar. Whole plants cultured from the leaf discs contained the genes for producing the glyphosate-resistant pathway for synthesis of the essential aromatic amino acids. Progeny from the plants were also glyphosate resistant.

It is quite possible that stress-resistance characteristics can be incorporated into plants but the characteristics must first be identified and understood. There are several factors that may make this approach more difficult than the examples given. The successes achieved using recombinant DNA techniques have been primarily limited to traits controlled by single gene manipulations. Multiple-gene interactions are common in plant stress responses and the engineering of a plasmid that possesses all the appropriate genes in the correct sequence may not be achievable. Finding a vector organism may be difficult also. For example, *Agrobacterium tumefaciens* colonizes only dicotyledenous plants and many of the economically important crops are monocotyledenous such as corn, wheat, oats, barley, and rice.

The use of a vector organism may be overcome by other techniques. Microinjection of DNA directly into a cell nucleus has been accomplished in animals and a technique referred to as electroporation has been used to transfer DNA into the protoplasts of corn. The protoplasts are subjected to an electrical current that temporarily opens the membrane pores and allows the DNA to penetrate the cell membrane.

A better understanding of the physiology, biochemistry, and development of stressed plants is needed if stress-tolerant plants are to be bioengineered.

STRESS PROTEINS AND TOLERANCE

Another example of an approach to answer questions regarding abiotic stress effects involves studies of the metabolic *de novo* synthesis of proteins. Most of the available information concerns proteins produced as a result of heat shock and proteins formed as a result of anaerobic conditions such as occur in flooded fields.

Heat shock proteins (hsp) were first discovered in the insect *Drosophila melanogaster* and later in plants. There are basically three different types known to occur in plants and they are grouped according to size as follows: group I, 68 to 104 kDa (found in bacteria, animals, and plants); group II, 20 to 33 kDa; group III, 15 to 18 kDa (found only in higher plants). In *Drosophila* and in soybean, but not in cereal grains, hsp are the only proteins synthesized during the stress period. Induction of hsp can be initiated by subjecting plants to temperatures approaching 40°C. In corn initial hsp production is as rapid as 20 min but a second set of proteins of a different size (52 to 62 kDa) are synthesized after 4 h. The total production of hsp decreases drastically after 4 h even though the heat stress is still present.

Anaerobic stress proteins (asp) are also expressed in plants subjected to anoxic conditions such as would occur in flooded fields. The asp produced in corn are different than the hsp produced as a result of heat stress except for one protein but both groups are biphasic. The asp are continuously produced for up to 72 h, after which the cells begin to die.

One of the asp produced in corn is alcohol dehydrogenase, the terminal enzyme in the pathway of fermentation. Fermentation occurs to a greater extent under anerobic conditions and is responsible for the survival of corn when flooded. Alcohol dehydrogenase recycles NAD^+ during fermentation and prevents the toxic acidosis that might occur from competing lactic and malic acid fermentations.

Other causes of stress protein formation have been identified and include high concentrations of heavy metals, ultraviolet light, and drought. Proteins produced by exposure to heavy metals may function as chelators of the metal causing detoxification. Proteins produced by exposure to UV light may be involved in the synthesis of the flavenoid pigments that absorb the potentially harmful irradiation. Little is known about drought induced proteins except that some of the hsp induced in soybean are also induced by drought stress and may be responsible for resistance to both stresses. The stress proteins may be involved in the synthesis of many of the nitrogenous compounds such as proline, polyamines, and betaines, which occur in drought-stressed plants.

Considering the role of membranes in sensing environmental stresses and triggering defense mechanisms, it is interesting to speculate that stress proteins may be involved. Plant growth regulators that improve tolerance to stress may affect stress protein induction also. The development of stress tolerant plants through the use of biotechnology is an attainable goal but one that requires the cooperative efforts of plant physiologists, biochemists, molecular biologists, and plant breeders.

DISCUSSION QUESTIONS

1. Discuss the various ways plants might be altered using PGRs to improve drought tolerance.

2. Describe some advantages and disadvantages of controlling transpiration by using PGRs that alter stomatal movement.

3. List some of the chemicals known to alter the morphology of plants and hence reduce transpiration rates.

4. What naturally occurring growth regulators have been used to change the temperature tolerance of plants?

5. Outline differences and similarities that occur physiologically between plants exposed to salt stress and to drought stress.

6. Development of tolerance by a plant to one stress often leads to the development of tolerance to other stresses. Give some examples and indicate the possible significance of such a relationship.

7. Outline several ways in which nutrient uptake efficiency might be enhanced by using PGRs.

8. In general terms describe how symbiotic and parasitic bacteria might be used to improve stress tolerance in higher plants.

9. Outline several limitations to the development of stress-tolerant plants using current genetic engineering techniques.

10. What are stress proteins and under what specific conditions are they produced?

11. Of what significance is the production of stress proteins in plants?

12. Outline some general characteristics of stress proteins.

REFERENCES

Carnahan, J. E., E. L. Jenner, and E. K. W. Wat. 1978. Prevention of ozone injury to plants by a new protectant chemical. *Phytopathology* 68:1225–1229.

Fenton, R. 1982. Evaluation of the possibilities for modifying stomatal movement. In J. S. McLaren (Ed.), *Chemical Manipulation of Crop Growth and Development*. Butterworth, Boston, pp. 19–37.

Jung, J., and W. Rudemacher. 1983. Plant growth regulating chemicals—Cereal grains. In L. G. Nickell (Ed.), *Plant Growth Regulating Chemicals*, Vol. 1. CRC, Boca Raton, FL, pp. 253–271.

Krizek, D. T., R. J. Giesbach, P. M. Hasegawa, R. A. Bressan, S. Handa, A. K. Handa, S. J. Stavarek, D. W. Rains, M. E. Daub, P. S. Carlson, and J. D. Lutz. 1984. Somatic cell genetics: Prospects for development of stress tolerance. *Hort. Sci.* 19:365–392.

Le Rudulier, D., A. R. Strom, A. M. Dandekar, L. T. Smith, and R. C. Valentine. 1984. Molecular biology of osmoregulation. *Science* 224:1064–1068.

Moses, P. B. 1987. Strange bedfellows. *Bioscience* 37:6–10.

Nickell, L. G. 1982. *Plant Growth Regulators*. Springer-Verlag, New York, Chaps. 13-15, pp. 51-59.

Rikin, A., D. Atsmon, and G. Gitler. 1979. Chilling injury in cotton (*Gossypium hirsutum* L): Prevention by abscisic acid. *Plant Cell Physiol*. 20:1537-1546.

Sachs, M. M., and T.-H. D. Ho. 1986. Alteration of gene expression during environmental stress in plants. *Annu. Rev. Plant Physiol*. 37:363-376.

Tomlinson, H., and S. Rich. 1973. Anti-sencescent compounds reduce injury and steroid changes in ozonated leaves and their chloroplasts. *Phytopathology* 63:903-906.

Valentine, R. C. 1984. Genetic engineering of salinity-tolerant plants. *Calif Agr*. 38:36-37.

Weaver, R. F. 1984. Changing life's genetic blueprint. *Natl. Geogr*. 166:818-847.

Weaver, R. J. 1972. *Plant Growth Substances in Agriculture*. Freeman, San Francisco, Chap. 11, pp. 371-433.

Wilson, W. C. 1983. The use of exogenous plant growth regulators on citrus. In L. G. Nickell (Ed.), *Plant Growth Regulating Chemicals*, Vol. 1. CRC, Boca Raton, FL, pp. 207-232.

GLOSSARY

Abiotic. Not involving or produced by organisms; used loosely for factors of the environment and their effects on plants producing stress and stress symptoms; physiogenic.

Acclimation. Nonheritable modification caused by exposure to new environmental conditions; organisms undergoing changes in physiological processes rendering them tolerant or resistant to stress; nonheritable alterations of phenotype.

Acropetal. Movement or translocation of solutes and hormones from the base to the tip of the plant.

Acute. An injurious event that usually involves necrosis such as fleck, scorch, or bifacial necrosis. In air pollution usage, injury caused by an air pollutant over a short time period; not chronic.

Adaptation. Processes of evolution causing a plant to be fit or adapted for the environment in which it thrives; if not successful, the species perishes.

Air pollutants. Gaseous or particulate substances released into the atmosphere in sufficient quantities or concentrations to cause injury to plant life.

Allelochemicals. Chemical compounds released from a living organism that affect other organisms in the vicinity.

Allelopathy. Biochemical interactions between living organisms; any direct or indirect effect by one plant on another through production of chemical compounds that escape into the environment.

Amphipathic. Phospholipids exhibiting polar charges on one end and nonpolar characteristics on the other end.

Amphipoteric. Phospholipid molecules having a polar hydrophilic head and a nonpolar hydrophobic tail.

Antiport. Permeases that transport two different substances in opposite directions simultaneously across membranes.

Apoplastic. Free space; upon entering the root on its way to the xylem, water moves through intercellular spaces and along cell walls following the pathway of least resistance.

Avoidance. Stress avoidance caused by the plant not coming to thermodynamic equilibrium with the stress; prevention or decrease in the penetration of a stress into tissues.

Basipetal. Movement or translocation of solutes and hormones from the tip to the base of the plant.

Biotic. Induced or caused by action of living organisms: a combining form used to complete words e.g., antibiotic, endobiotic, exobiotic, of or relating to life.

Boundary layer. A layer of undisturbed air or solvent next to a surface; commonly the layer of air at the surface of a leaf.

C3 cycle. Carbon assimilition that results in formation of three-carbon compounds such as phosphoglycerate; the Calvin cycle; C3 plants are those having the cycle.

C4 cycle. Carbon assimilation that results in the formation of four-carbon compounds such as malate or aspartate; plants having the cycle are sometimes called C4 plants.

Calcicole. A plant that is able to grow and develop on calcareous soils.

Calcifuge. A plant that is injured, killed, or inhibited on calcareous soils.

CAM. See crassulacean acid metabolism.

Casparian strip. The heavily suberized transverse and radial walls of the endodermis; water and solutes are unable to leak past the endodermis through intercellular spaces and must pass through the cytoplasm.

Chelate. A metal ion plus a ligand; two important metal chelates found in nature are chlorophyll and heme; chelates play an important role in metabolism and in detoxification of heavy metals.

Chlorosis. Disappearance of chlorophyll as a result of altered metabolism brought on by disease or stress.

Chronic. An injurious event that occurs over time; exposure to an unfavorable environmental factor for a prolonged time, resulting in injury; as opposed to acute in air pollution injury.

Coordination number. The number of positions within each central atom's (metal) coordination sphere that can be occupied by ligands, irrespective of whether the ligands are neutral or charged. The maximal number of functional groups with which a metal will complex.

Crassulacean acid metabolism. CAM metabolism occurs in species of plants that have open stomates at night and fix CO_2 into organic acids such as malic acid, which is stored until daylight occurs; decarboxylation occurs in daylight, releasing CO_2; CAM plants are able to effectively photosynsthesize without open stomates during periods favoring high transpiration rates.

Critical concentration. That concentration of a nutrient in plant tissue just at the level giving optimal growth.

Cryoprotectant. A chemical compound whose presence prevents injury during freezing; may occur naturally or be applied.

Cuticle. The waxy deposit on the surface of the plant outside the epidermis.

Cuticular resistance. A measure of the reduction in diffusion through the plant tissues as a result of the presence of the cuticle; the reciprocal of conductance; cuticle may vary in thickness, porosity, and composition rendering more or less resistance to diffusion of gases such as water vapor and carbon dioxide.

Cuticular transpiration. Outward diffusion of water vapor through the cuticle.

Darcy's laws. The first law is for isothermal, steady state liquid water flow in a porous medium; the second law states that water will flow across an air–water interface only when the pressure potential is sufficiently greater than zero to overcome the surface tension of the fluid.

Dehydration avoidance. Prevention of loss of water by closing of stomates resulting in the maintenance of turgor during periods favoring high rates of transpiration.

Dehydration postponement. Reduction of transpiration or increase of water absorption because of morphological characteristics of drought-stressed plants.

Dehydration tolerance. The ability of plants to withstand injury at the protoplasmic level when plants are under drought stress.

Desiccation tolerance. Protoplasmic resistance to injury under extreme water-stress conditions; upon rehydration, respiration and photosynthesis begin within a relatively short time.

Disease. A condition of the living animal or plant body or one of its parts that impairs the performance of a vital function; impairment or modification of the performance of a vital function or functions; response to environmental factors or inherent genetic defects.

Dormancy. A state of inactive growth; typically used in reference to buds and seeds; important in resistance to stresses.

Dose. Exposure to a measured concentration of a toxicant for a known duration of time.

Drought. A meteorological term for prolonged periods of reduced precipitation.

Drought resistance. Innate mechanisms by which injury is prevented or minimized during drought stress; some scientists prefer to use the term drought tolerance instead of resistance.

Drought tolerance. Result of postponement of dehydration (see dehydration postponement) or because of innate dehydration tolerance characteristics.

Electrochemical potential. A potential across a membrane composed of two components, a concentration component and an electrical component; final direction of diffusion of a solute is determined by the component that has the steepest gradient across the membrane.

Entropy. A measure of the degree to which a system is removed from equilibrium; as the degree of randomness increases in a system, entropy increases.

Escape. Because of timing sequence in development and growth, some plants are not subjected to a stress; life cycle may be completed during a period or season when a particular stress does not occur.

Euhalophytes. Halophytes that grow in conditions of extreme salinity.

Euxerophytes. Plants that grow under dry conditions, such as in deserts.

Facultative halophytes. Plants that acclimate to saline conditions.

Flux. Volume flow across a unit area per unit time; usually given in mol cm^{-2} s^1 for example.

Freezing avoidance. Effects of freezing are excluded from protoplasm.

Frost hardiness. Frost tolerance; may be induced by gradual exposure to lowered temperatures.

Frost resistance. Frost tolerance.

Frost tolerance. Injury does not occur at temperatures of -1 to $-3°C$

Gibbs free energy. Expressed as $G = E + PV - TS$; the energy isothermally (constant temperature) available for conversion to work; always measure ΔG where ΔG is given in energy/mole, E = internal energy, PV = pressure-volume constant, T = absolute temperature, and S = entropy.

Gibberellins. Growth regulators that cause cell enlargement; promote shoot growth.

Glycophytes. Plants that are sensitive to relatively high soil salt concentrations.

Growth inhibitors. Organic chemicals that slow growth; may be naturally occurring or applied.

Growth regulators. Organic chemicals that promote growth when present in relatively low concentrations and may inhibit growth at higher concentrations.

Halophytes. Plants that are able to grow in the presence of high soil salt concentration; (see also euhalophytes, facultative halophytes, obligate halophytes).

Hardening. Acclimation; increased tolerance of stress; may be caused by gradual exposure to stress; changes may be phenotypic or occur at the protoplasmic level.

Hardiness. Ability to withstand or tolerate stess conditions.

Health. The condition of being sound in body; freedom from disease.

Heat tolerance. High-temperature tolerance; means of dissipating high-temperature effects are present or physiological changes occur that prevent injury.

Homeostasis. A relatively stable state of equilibrium or tendency toward such a state between the different but interdependent elements or groups of elements of an organism or group.

Homogeneous nucleation. Supercooled water freezes instantaneously when a nucleator is introduced but at a temperature of $-38.1°C$ water in small droplets freezes automatically.

Hormone. A specific organic product produced in one part of a plant or animal and transported to another part where it is effective in small amounts in controlling or stimulating processes.

Hydrogen bonding. The three elements N, O, F, are the three most electronegative elements; because of this, they apparently attract the electrons of the covalent bond to such an extent as to leave the H nucleus almost completely exposed; H is the smallest of all atoms, therefore it is possible for it to come very close to an electronegative atom of another molecule; the resulting electrostatic attraction, while weak compared to a covalent bond, is much stronger than the attraction between dipoles or ordinary Van der Waal's forces and for this reason is called a hydrogen bond.

Hydrophytes. Plants that live in water or in virtually water-saturated soils.

Ice nucleators. Materials that cause freezing of water when it is supercooled; ice crystals may serve when introduced into supercooled water.

Injury. A change in metabolism that results in altered functions and/or appearance with deleterious effects.

Ionic bond. The strong electrostatic attraction between two oppositely charged ions.

Ionization constant. Written as K_a or K_b; an indication of the constant strength of an acid or base; the greater the degree to which the acid or base ionizes at a given concentration the stronger it is, and the larger is the value of its ionization constant.

Lateral phase separation. Separation of phospholipids in a heterogeneous mixture into more homogeneous groups in membranes when the temperature is reduced.

Lewis acid. Any substance capable of acting as an electron pair acceptor: yields a hydrogen ion in solution.

Lewis base. Any substance possessing a pair of electrons available for sharing with a proton to form a covalent bond; yields a hydroxyl ion in solution.

Ligand. Can be either neutral molecules or negative ions, both of which possess unshared electron pairs and are capable of functioning as electron donors, that is, as Lewis bases; each molecule or ion possesses, therefore, at least one donor center capable of occupying a position in the coordination sphere of the metal atom; one donor center is described as unidentate while a ligand with two or more donor centers is said to be multidentate; multidentate ligands, such as EDTA, are commonly called chelating agents because of the manner in which they enclose the central atom in a pincer-like manner reminiscent of the action of a crab's claw (Greek: chele); the complexes are called chelates.

Light harvesting complex. Complex of pigments that intercept light impinging on leaves; chlorophylls plus other pigments.

Light saturation. The light intensity at which no further increase in intensity can cause an increase in photosynthesis.

Liquid crystalline. Fluid state of pure phospholipids; related to phase transitions in membranes; caused by temperature.

Luxury consumption. Absorption of mineral nutrients above and beyond those that cause increases in growth or yield.

Matric potential. (Ψm); The term is used to account for all the forces of capillarity, adsorption, and hydration that cause imbibition or holding of water in a matrix of any sort.

Membrane carriers. Molecules or compounds that assist materials to cross membranes; associated with transport of charged particles such as ions; many are proteins.

Membrane fluidity. Changes with temperature and amount of cross linkage.

Membrane transition. Change in the physical state of membrane components as temperature is lowered; change from liquid crystalline state to gel structure.

Mesophytes. Plants that thrive best with a moderate water supply.

Metalloenzymes. Enzymes that contain a transition metal as the prosthetic group.

Mycorrhizae. A symbiotic association between a fungus and usually the root of a higher plant; the fungus normally makes available to the root forms of elements (especially N and P) that would otherwise be unavailable; mycorrhizal associations occur throughout the plant kingdom.

Natrophilic. "Sodium loving"; refers to plants that grow in sodium-rich soils.

Necrosis. Death.

Noninfectious disease. A disease that is caused by an environmental factor, not by a pathogen; caused by a physiogen; a physiogenic disease.

Nutrient balance. The total number, on an equivalent basis, of cations and anions absorbed by a plant; plants under uniform environmental conditions tend to take in a constant number of cations and anions; since the total number of cations and anions absorbed remains constant, if K^+ in the plant were increased, Ca^{2+} and Mg^{2+} would tend to decrease and vise-versa.

Obligate halophytes. halophytes that are able to grow only in saline soils.

Obligate shade plants. Plants that grow well in reduced light intensities but are injured when placed in relatively high light intensity.

Oligohalophytes. Halophytes that grow in moderately saline soils.

Osmoregulation. The control of plant functions by the regulation of water potentials across differentially permeable barriers or membranes; water diffusing down a potential gradient is losing energy and can be made to do work; an example would be stomatal control by the regulation of the turgidity of guard cells.

Osmoprotectants. Organic solutes produced as a result of salt stress.

Osmotic adaptation. Osmotic adjustment.

Osmotic adjustment. Increase in organic solute content as a result of stress; lowers osmotic potential of cell sap.

Osmotic potential. ($\Psi\pi$) The component of water potential that becomes increasingly negative with the addition of solute.

Pathogen. A disease-initiating biotic entity; usually restricted to organismic causes e.g., bacteria, fungi, viruses.

Pathogenesis. The origination and development of a pathogenic disease.

Permanent wilting percentage. (PWP) A moisture content at which the soil is incapable of maintaining succulent plants in an unwilted condition even though transpirational loss is negligibly low; approximately equivalent to -1.5 MPa.

Permeability. (L) a property of a membrane describing material movement across the membrane.

Phase separation. As water in tissues freezes, the solid phase (ice) is separated from liquid, unfrozen water.

Phase transition. Change from liquid to solid as from liquid water to solid ice; (see lateral phase transition in membranes).

Photophosphorylation. The synthesis of ATP in the illuminated chloroplasts.

Photo-inhibition. Light-reduced activity such as growth or an enzymatic reaction.

Photo-morphogenetic. Ability to change morphology as a result of exposure to light.

Physiogen. A disease-initiating abiotic entity or event; caused by physical or chemical factors in the environment.

Physiogenesis. The origination and development of a physiogenic disease.

Plasmolysis. The condition of a cell in which the turgor has been reduced by removal of water by osmosis; the cytoplasm may separate from the wall of the cell.

Poikilotherms. Organisms that attain the temperature of the environment while still alive; plants and some animals.

Recombinant DNA. Inserting or replacing a fragment of a DNA molecule thus changing the genome of an organism.

Resistance. Tolerance or avoidance; the ability of an organism to withstand or overcome partially or completely the injury caused by a physiogenic or a pathogenic factor; the means of resisting, fighting against, or counteracting the injurious effects of environment.

Rhizosphere. A volume of soil near the root surface that is affected by the presence of the root and the chemical and physical interactions that occur between the root and its immediate environment; the root's sphere of influence.

Relative humidity. (rh) the percentage of water vapor present in the atmosphere relative to saturation (100%) at the existing temperature.

Saturated flow. Water flow in a soil in which all pores are filled with water; the hydraulic conductivity K in a specific saturated soil is at its maximum and is constant.

Solar constant. Intensity of unfiltered sunlight entering the ionosphere; $1.39 \text{ kW}/\text{m}^2$.

Stability. As applied to chelates, thermodynamically evaluates the feasibility and predictability of resistance or destruction of complexes under various conditions.

Stele. The central cylinder inside the cortex of roots and stems of vascular plants; composed of tissues inside the endodermis.

Stomatal conductance. Rate per unit area at which gases diffuse through the stomates.

Stomatal resistance. The reciprocal of stomatal conductance.

Strain. An injury of a body part or organ resulting from excessive tension, force, influence, or factors causing excessive physical tension.

Stress. A physical or chemical factor that causes bodily tension and may be a factor in disease causation; a state or condition caused by factors that tend to alter an equilibrium; the state or condition of strain; expressed quantitatively in units of force per unit area.

Stress proteins. Proteins induced to form as a result of the stress; for example, high temperature-induced proteins.

Symptom. Evidence of disease or physical disturbance; something that indicates the presence of a bodily disorder; an evident reaction of a plant to a pathogen or physiogen.

Symplastic. Refers to the pathway that water and solutes may follow upon entering the root if they cross the plasmalemma and move through the interconnecting plasmodesmata from cell to cell.

Symport. Permeases that transport two different molecules in the same direction simultaneously.

Syndrome. A group of factors, signs, and symptoms that collectively indicate a disease or disorder.

Synergistic. Two substances separately produce an effect but when used together cause a greater effect than the sum of the two acting separately.

Tolerance. Ability of a plant to survive a stress with little or no injury; the relative capacity of an organism to grow, thrive, or survive when subjected to an unfavorable environmental factor; the capacity to resist or sustain the effects of a disease or stress without dying or suffering irreparable injury.

Turgor. (Ψp) the pressure exerted on the cell contents by the walls of a turgid cell; or conversely the pressure exerted by the water in the vacuole against the cytoplasm and the cell wall.

Uniport. Permease that transports one substance in one direction across a membrane.

Unsaturation. The number of double bonds of a molecule.

Van der Waal's forces. Forces that arise from attractive and repulsive forces between molecules; the repulsive forces diminish with distance more rapidly than the attractive ones, leading to a net attraction at short distances; forces arise from the attraction of the positively charged nuclei of one molecule and the negatively charged electrons of another; the forces are relatively weak.

van't Hoff relationship. ($\Psi\pi = -miRT$) relates osmotic potential to gas laws.

Water potential. (Ψw) the difference between the activity of water molecules in pure distilled water at atmospheric pressure and 30°C and activity of water molecules in any other system; the addition of solutes to water decreases the water potential.

Water stress. Usually reserved for short term situations in which transpiration rate exceeds absorption rate of water; may be caused by a number of factors.

Wilting. Reduction in cell turgor that reduces the rigidity of a plant; result of water stress.

Xerophytes. Plants that are able to resist drought and live in very dry habitats; often the plants have thick cuticles, water storage tissue and other morphological features that conserve water.

Zero stress. In concept that level of exposure to an environmental factor that leads neither to injury nor to reduction in growth, yield, or value.

INDEX

ABA, *see* Abscisic acid
Abies balsamea, 147
Abscisic acid (ABA):
  and disease, 164
  and drought, 41, 142, 145, 147–151,
    165, 166, 172–174
  and light, 111, 112, 158, 160, 161
  and nutrition, 156, 167
  and salt, 175
  and temperature, 152, 153, 165, 174, 182
Abscission, 36, 112, 147, 148, 151, 156,
  164, 165
Absorption:
  of allelochemicals, 124
  of aluminum, 84
  of CO_2, 110
  of ions, 74, 151
  of light, 106, 107, 109, 110, 111
  of manganese, 86
  of nutrients, 7, 21, 56, 73, 82, 90, 142, 189
  of radiation, 36
  of salt, 96
  of water, 5, 6, 8–12, 46, 55, 98, 172, 186,
    194
ACC, *see* Aminocyclopropanecarboxylate
  (ACC)

Acclimation to temperature, 46, 60–64, 67,
  68, 70, 152, 169, 188
Acer negundo, 174
Acidic phospholipids, 138
Acidosis, 180
Acid phosphatase, 88
Active transport, 46, 55, 124, 134, 135,
  139, 142, 143
Adaptation:
  defined, 4, 183
  to drought, 29, 36, 42, 43, 150, 166, 168
  morphological, 36, 42, 103, 113
  to nutrient stress, 89, 91, 92
  osmotic, 29
  physiological, 43
  to salinity, 100, 101
  to shade, 109, 114
  to temperature, 63, 67, 69, 70
Adenosine triphosphate (ATP), 16, 19, 134,
  135, 142, 191
Adventitious roots, 22, 24, 162
Aerenchyma, 22, 24, 95
Agrobacterium tumefaciens, 177, 178, 179
Air pollutants, 1, 75, 177, 183, 185
Alanine, 20
Alfalfa, 85, 97, 101, 118, 126

Alkenylsuccinic acid, 173
Alkylene diamines, 174
Allelochemical action, 124
Allelochemical classification, 121
Allelochemicals, 1, 184
Allelochemical sources, 120, 121
Allelopathy, 118, 119, 120, 123, 124, 184
Aluminum, 74, 84
Aluminum accumulator plants, 85
Aluminum excluder plants, 85
Aluminum tolerance, 85, 86, 88
Aluminum toxicity, 90, 94
Amines, 130
Amino acid metabolism, 23, 25, 56, 69, 78, 86, 92, 130, 135, 154, 167
Amino acids as osmotica, 32, 34, 35, 55, 97
Aminocyclopropanecarboxylate (ACC), 146, 148, 153, 154, 156, 167
2-amino-6-methyl benzoic acid, 175
Ammonium, 85, 154
Ammonium (5-hydroxycarvacryl) trimethyl-chloride piperidine hydrochloride (AMO-1618), 173
Ammonium nitrate, 154
Ammonium sulphate, 154
AMO-1618, 173
Amphipathic compounds, 133, 184
Amphipoteric compounds, 140, 184
Anaerobic conditions, 21, 24, 148
Anaerobic stress proteins (asp), 179, 180
Anticytokinin, 173
Antifungal quinones, 164
Antimetabolites, 46
Antiport permeases, 134, 184
Antisenescence chemicals, 177
Antitranspirant chemicals, 172
Apical dominance, 112, 150
Apple trees, 54, 118, 127, 146, 154, 157, 174
Aqueous pores, 141
Arid, 121, 124
Ascorbic acid oxidase, 77
ATPases, 134, 135, 139
ATP synthetase, 110
Atriplex patula, 98
Ausdauer unit, 38
Auxin, *see* Indoleacetic acid (IAA)
Avena sativum, 125
Azalea, 85, 174

BA, *see* Benzyladenine (BA)
Bakanae disease, 164
Barley, 21, 24, 26, 40, 41, 43, 85, 88, 91, 97, 98, 100, 148, 175, 179
Barrel cactus, 35
Beans, 15, 24, 41, 42, 152, 159, 161, 162
    broad, 147, 148
    dwarf, 148
    jack, 15
    kidney, 96, 101
    lima, 175
    mung, 156
    pinto, 177
    pole, 175
    snap, 175
    white, 177
Beets, 152, 162
Benzimadazole, 177
Benzoic acid, 124, 175
Benzyladenine (BA), 174, 177
Beryllium, 83
Betaine, 21, 24, 41, 175, 180
Bilayer of membrane, 129, 130, 132–135, 138, 139, 140
Biotechnology, 171, 177, 180
Birch trees, 154
Black walnut trees, 118, 119, 127
Blossom end rot, 139
Boron, 74, 113, 156, 167
Box elder, 173
Breeding:
    for allelopathy, 118, 124, 125, 126
    for drought tolerance, 39, 41, 42, 43
    for irradiation tolerance, 113
    for nutrient tolerance, 88, 90, 176
    for salt tolerance, 94, 97, 98, 100
    for temperature tolerance, 46, 57, 58, 68
Broad bean, 147, 148
Bulliform cells, 36
Bundle sheath cells, 18, 20

Cabbage, 152, 173
Cadmium, 83
Calcicole plants, 94
Calcifuge plants, 94
Calcium (Ca), 74, 79, 84, 88, 89, 94, 113, 156, 167, 168, 176
    in membranes, 138, 139, 141, 143, 144, 156

CAM metabolism, *see* Crassulacean acid metabolism (CAM)
Campesterol, 131
Canavalia ensiformis, 15
Capillary action, 73
Carbon dioxide, 18, 19, 20, 25
   in troposphere, 104
Carbon dioxide compensation concentration, 17
Carbon dioxide fixation, 16, 110
Carbon dioxide light effects, 110
Carbon metabolism:
   C3, 16–20
   C4, 16–20, 103
   CAM, 19, 25, 97
Carbon monoxide, 77
Carnation, 175
Carriers, *see* Membranes
Casparian strip, 9, 185
Catalase, 78
CBBP (2,4 dichlorobenzyltributyl-phosphonium chloride), 173
CCC, *see* Chlorocholine chloride (CCC)
Cereals, 89, 97, 175
Chelation, 83, 85, 87, 88, 90
Chenopodium album, 98
Chilling injury, 46, 47, 66, 68
Chilling temperature, 136
Chilling tolerance, 52
Chitinase, 164
Chloride ions, 95, 99
Chlorocholine chloride (CCC), 173–176
Chlorophyll, 15, 79, 107–111, 114, 162
Chlorophyll photodestruction, 110
Chloroplast membrane, 15, 63, 109, 111, 129, 141
Chloroplasts, 18, 20, 61, 68, 107, 109, 150, 177, 182
Chloroplast thylakoids, 47, 63, 117, 135
5-chloro-4-quinoline carboxylic acid, 175
Chlorosis, 75, 78, 79, 86, 124, 162
Cholesterol, 52, 131, 136, 137, 139, 144
Choline, 173, 175
Chromatin, 89
Chromophores, 111, 114, 157
Chromosomes, 111
Chrysanthemum, 148
Cinnamic acid, 124
Citrate, 79

Citrus trees, 45, 70, 174, 175, 182
CK, *see* Cytokinin (CK)
Cobalt, 83, 156
Cold hardening, 50, 52, 54, 60, 64, 65, 67–70
Coleoptile, 156, 158, 159
Compartmentation of cells, 97, 150, 154
Coordination number, 88, 185
Copper, 77, 78, 79, 83, 86, 89–92, 156
Corn, 15, 20, 22, 24, 26, 41, 97, 101, 151, 152, 153, 179, 180
Cortical cells, 84, 151
Cortical tissues, 9, 154
Cotton, 40, 41, 86, 92, 147, 174, 182
Coumarins, 124
Cranberry, 85
Crassulacean acid metabolism (CAM), 19, 25, 97, 185
Cressa, 95
Critical concentration, 72, 80, 90, 185
Critical drought avoidance, 38
Critical temperature, 47
Cryoprotectants, 45, 185
Cucumber, 125, 126, 153, 156, 157, 164, 169
Cucumber mosaic virus, 164
Cucumis sativus, 125
Cuticle, 106, 111, 185
Cuticular diffusive resistance, 35, 36, 38, 40, 41, 185
Cuticular transpiration, 38, 185
4-cyclopentylamino-2-methylthiopyrolo(2,3-*d*)-pyrimidine, 173
Cytochrome *b*559, 110
Cytochrome *b*563, 110
Cytochrome *f,* 110
Cytochrome oxidase, 77, 78
Cytokinin (CK):
   in disease, 162, 164, 167
   in drought (water relations), 142, 145, 148, 151, 165, 175
   in irradiation stress, 157, 158, 161, 169
   in nutrient stress, 154–157, 167, 168
   in temperature stress, 153, 154, 174
Cytoplasmic membrane, 157

Darcy's law, 7, 186
Deciduous forest trees, 54

Deciduous fruit trees, 45, 54
Deficiency, *see* Nutrient
Dehydration, 140, 141, 143
  avoidance, 27, 28, 186
  postponement, 27, 28, 186
  tolerance, 27, 28, 38, 186
Deoxyribonucleic acid (DNA), 111, 112, 114
Deprotonated membrane system, 139
Desaturase enzymes, 137, 143
Desiccation tolerance, 37, 39, 186
Detoxification, 85
Diagnostic tool, 79
2,4-dichlorobenzyl tributylphosphonium chloride (CBBP), 173
2,4-dichlorophenoxyacetic acid (2,4-D), 176
Differential tolerance, 85
Diffusive resistance, 39
Divalent anions, 99
DNA, *see* Deoxyribonucleic acid (DNA)
Dormancy, 151, 152, 186
Downy mildew, 162
Drought, 1, 3, 10, 11, 24, 186
  avoidance, 28, 38, 41, 42, 43
  escape, 28, 34, 40
  and phytohormones, 146, 148, 150, 151, 165
  resistance, 27, 28, 39, 40, 41, 43, 69, 186
  stress, 5, 6, 9, 12, 14–43, 172, 175, 176, 178, 180, 181
  tolerance, 28–43, 58, 140, 156, 172, 176, 180, 186
Duckweed, 161
Dwarf bean, 148

Ectomycorrhizae, 87
EDU (N-[2-(-oxo-1-imidazolimdinyl), ethyl]-N′-phenylurea), 177
Effectors in transcription, 142
Eggplant, 142
Electrochemical equilibrium, 134
Electrochemical potential, 186
Electrolyte leakage, 97
Electromagnetic radiation, 111
Electron transport, 135, 141
Electroporation of membranes, 179
Endodermis, 9

Endoplasmic reticulum, 129
Entrappers of aluminum, 85
Enzymes, 20, 38. *See also specific enzymes*
Ephemeral plants, 34
Epidermal cells, 77
Epidermal transfer cells, 89
Epidermis, 91, 112
Epinasty, 11, 21, 162, 163
Escherichia coli, 175
Essential nutrient element, 71, 73, 77, 81, 83, 84, 85, 88
ETH, *see* Ethylene (ETH)
Ethephon, 22
Ethylene (ETH):
  in disease, 162–165, 168
  in drought (water relations), 145–151
  in flooding, 21, 22, 24, 25
  in irradiation stress, 113, 154–162, 167
  in nutrient stress, 154–157, 167
  in temperature stress, 152–154, 165, 167, 169
Etiolated tissues, 157, 161
Euhalophytes, 94, 187
Euxerophytes, 187
Extracellular ice, 48, 52, 65

Facilitated diffusion, 134, 143
Facultative halophytes, 94, 187
Far-red light (FR), 108, 109, 157–160
Fatty acids, 173
Ferredoxin, 110
Flacca tomato mutant, 150
Flavanoids, 112
Flavoproteins, 111
Flooding, 1, 21, 22, 24, 25
Flooding injury symptoms, 22
Fluorescence, 19
Fluorescent chemicals, 125
Flux, 187
Frankenia, 95
Fraxinus pennsylvanica, 22
Free radicals, 110, 112, 113
Freeze–thaw cycles, 50, 51, 52
Freezing, tolerance, 45, 49, 52, 54, 69
Freezing avoidance, 45, 187
Freezing dehydration, 47, 48, 50, 52
Freezing injury, 46, 47, 49, 51, 52, 53, 66–70
Freezing point, 48, 66

Freezing process, 47, 49, 52, 53
Freezing rate, 65
Freezing tests, 58
Frost hardiness, 45, 54, 66, 187
Frost injury, 54, 55
Frost injury resistance, 50, 54, 187
Frost tolerance, 54, 55, 187
Fruits, 77, 82, 96
Fruit trees, 155, 175
Fungal mantle, 87

GA, *see* Gibberellins (GA)
Galactolipids, 130
Gall formation, 113, 164
Gamma rays, 112
Gel state, 136
Genetic engineering, 171, 177, 178, 181
Geotropic responses, 113
Geranium, 175
Gibberella fujikuroi, 164
Gibberellins (GA):
    in disease, 164, 165, 168
    in drought (water relations), 146, 165
    in irradiation stress, 158–161, 166
    in membranes, 142
    in nutrient stress, 176
    in temperature stress, 153, 174
Gibbs free energy, 49, 187
Glaux, 95
Glucose, 134
Glycerol, 130, 139
Glycine, 20
Glycolate pathway, 20
Glycophytes, 94, 95, 187
Glyphosate, 178
Gossypium hirsutum, 182
Grana, chloroplast, 107, 111
Grape, 174, 175
Grasses, 78, 85, 98, 118, 158
Growth, effects:
    of disease, 163, 164
    of drought, 6, 8, 10, 11, 13–16, 22, 23, 25, 26, 146, 150, 151, 172, 173
    of irradiation stress, 103, 108–112, 156, 158–161, 166
    of nutrient stress, 154, 156, 157, 168, 176
    of temperature stress, 46, 53, 56, 63, 64, 66, 68, 69, 152, 153, 166–169
Growth inhibitors, 187

Growth promoters, 121
Growth regulators (PGR), 22, 112, 171, 180. 181, 182, 187
Growth retardants, 172, 173
Guard cells, 150, 151, 172

Halophytes, 94, 95, 97, 100, 101, 102, 187
Hardened plants, 50, 52, 65, 67, 69
Hardening process, 60, 64, 65, 67, 68, 70, 188
Hardiness, 45, 46, 54, 65, 66, 67, 69, 70, 188
Heat shock proteins, 179
Heat tolerance, 39, 58, 188
Heliotropism of leaves, 111
Hemes, 78
High intensity light, 79, 98, 101, 110
High temperature acclimation, 63
High temperature injury, 46, 58, 59
High temperature stress, 56, 58, 121, 165
High temperature tolerance, 58, 59, 63, 174
Hill reaction, 19, 20
Homeostasis, carbon dioxide, 20
Homogeneous nucleation temperature, 48, 188
Hormones, *see specific phytohormones*
Hydration, 132, 138–141, 143
Hydration energy, 88
Hydrogen bonding, 139, 188
Hydrogen ions, 85, 89, 135, 156
Hydrogen peroxide, 164
Hydrolases, 15
Hydrophilic heads (of phospholipids), 133, 140
Hydrophobic forces, 49, 63
Hydrophobic tails (of phospholipids), 133, 140
Hydrophytes, 188
Hydrostatic feedback, 35
Hydrostatic pressure, 6
Hydroxides, iron, 78
5-hydroxy-alpha-naphthaquinone, 119
Hydroxylases, 163
Hyperplasia, 162
Hypertrophy, 22

IAA, *see* Indoleacetic acid (IAA)
Ice nucleators, 46, 48, 188
Incident radiant energy, 104, 106

Indoleacetic acid (IAA):
    in disease, 162, 163, 164
    in drought (water relations), 146–148, 165
    in irradiation stress, 112, 158–161, 166
    in membranes, 142
    in nutrient stress, 154–157
Indoleacetic acid oxidase, 146, 156, 159, 163
Infrared radiation, 106
Insects, 1, 40, 125, 162, 164, 165
Interveinal chlorosis, 79
Intracellular ice, 48, 52
Intracellular nucleation, 54
Intracellular water, 48
Ionic bond, 188
Ion interactions, 78, 84, 132
Ionization constant, 189
Ionizing radiation, 112, 114
Ion stress, 94, 95, 99
Iron (Fe), 74, 77, 78, 79, 86, 91, 156
Iron efficiency, 89
Isozymes, 38, 52

Jackbean, 15
Juglans, 119, 126, 127
Juglone, 118, 119, 126

Kidney bean, 96
Kinetic motion, 132, 133

Lactose, 134
Lateral phase separation, 136, 138, 139, 143, 189
Leaching, 118, 120, 127, 176
Leaf abscission, 147, 150, 151, 164
Leaf area, 34, 35, 36
Leaf expansion, 159
Leaf miners, 162
Leaf puckering, 86
Leaf rolling, 36, 161
Leaf senescence, 148, 162
Leaf water potential, 29, 32, 33, 36, 41, 148, 150
Leaf yellowing, 75, 79, 84
Leghemoglobin, 78
Legumes, 158
Lemna, 161

Lesions, 75, 111
    membrane, 69
    metabolic, 75
    necrotic, 59
    photochemical, 111
Lettuce, 146, 152, 160, 175
Lewis acid, 78, 88, 189
Lewis base, 88, 189
Ligands, 77, 78, 86, 87, 88, 189
Light:
    membrane effects, 132, 135, 136, 141, 142
    and phytohormones, 150, 157–161, 166, 167, 169
Light compensation point, 106, 113
Light deficit, 109, 113
Light harvesting complex (LHC), 108, 189
Light intensity, 28, 79, 98, 101, 106–110, 113, 114
Light quality, 103–106, 108, 115
Light reactions, 110
Light saturation, 110, 189
Light stress, 104, 111
Lima beans, 175
Lipases, 175
Lipid layers, 49
Lipid peroxidases, 52
Lipid phase change, 46, 66
Lipids, *see* Membrane lipids
Liquid-crystalline, 47, 136, 189
Long days, 158
Luxury consumption, 72, 90, 189

Magnesium (Mg), 84, 138, 139, 144
Malate decarboxylation, 19
Maleic hydrazide, 174
Malformations, 75
Malfunctions, 75
Malic acid, 17, 18, 19
Maltose, 134
Manganese (Mn), 77, 78, 79, 84, 89, 90, 92
Manganese toxicity, 85, 86
Mangroves, 85
Mass flow, 73, 74
Matric potential, 6, 7, 13, 189
Megapascals (MPa), 7, 12
Melting point depression, 48

Membrane lipids:
  temperature effects, 53, 59, 63, 64, 70
  UV effects, 111
Membrane permeability, 121, 141, 163, 191
  cation effects, 138, 156, 166
  PGR effects, 173
  phytohormone effects, 142, 143, 151, 163
  temperature effects, 132, 136
Membrane proteins, 47, 52, 66, 88
Membranes, 6, 12, 15, 16, 111, 173, 180
  carriers in, 133, 134, 135, 173, 190
  of chloroplasts, 109, 111
  fluidity of, 46, 132–139, 143, 144, 190
  lipids in, 129, 130, 132, 133, 135–140,
    142, 144, 173, 180
  of organelles, 12, 129, 130, 133, 135, 141,
    154
  of thylakoids, 47, 63, 107, 135
Membrane transition, 47, 52, 190
Meristems, 2, 13, 14
Mesembryanthemum crystellinum, 19
Mesophyll chloroplasts, 150
Mesophytic plants, 150, 190
Metalloenzymes, 88, 190
3-methyl-7-pentylamino-pyrazoli (4, 3-*d*)-
    pyrimidine, 173
Mevalonic acid, 122
Microinjection, 179
Micrometabolic nutrients, 77, 83, 85
Mild stress, 2, 12, 13, 15, 17, 20, 27
Mineral nutrient, *see* Nutrient
Mites, 164
Mitochondria, 47, 57, 70, 89, 129, 135, 141,
    157
Moderate deficiency, 72
Moderate salinity, 94
Moderate stress, 12
Moderate temperature, 98
Molybdenum, 154
Monodentate ligands, 86
Monoribosomes, 21
Monovalent cations, 99, 138, 139, 140, 144
Morphogenesis, 145, 157, 158, 165, 167,
    169
Mucigel, 9
Mucilage, 9
Mulberry, 174, 175
Mung bean, 156

Muskmelon, 152
Mutagenizing techniques, 178
Mutations, 111, 112, 113
Mycorrhizae, 87, 91, 176, 190

Naphthalene acetic acid (NAA), 175
Natrophilic plants, 94, 190
Natrophobic plants, 94
Necrosis, 46, 59, 75, 78, 86, 156, 190
Net assimilation rate, 125
Neutral phospholipids, 138
Neutrons, 112
Nickel, 83
Nicotine adenine dinucleotide
    (NADH+H), 135
Nitrate, 20, 79, 85, 99, 154
Nitrate reductase, 20, 111, 154
Nitrifying bacteria, 176
Nitrogen, 78, 82, 85, 90, 91, 154, 155, 156,
    166, 167, 168, 176
Nitrogenase, 78
Noninfectious disease, 190
Nonpolar phospholipids, 133, 134, 140, 143
N-[2-(oxo-1-imidazolidinyl),
    ethyl]-*N'*-phenylurea, 177
Nucleic acids, 46, 111, 114
Nutrient:
  absorption, 73, 74, 86, 87, 90, 176, 181
  balance, 80, 81, 90, 190
  critical concentration, 80, 81, 90
  deficiency, 75, 80, 83
  interaction, 78, 84, 86
  stress, 72, 89, 90
  stress tolerance, 3, 85, 90
  toxicity, 75, 80, 83
  translocation, 20, 71, 73, 74, 78, 82, 86

Oats, 125, 175, 179
Obligate halophytes, 94, 190
Obligate parasites, 162
Obligate shade plants, 107, 190
Okra, 152
Oligohalophytes, 94, 190
Orange trees, 175
Organelle membranes, 12, 129, 130, 133,
    135, 141, 154
Organic acids, 16, 20, 85
Organic solutes, 97

Osmoprotectants, 175, 178, 191
Osmoregulation, 29, 181, 190
Osmotic adaptation, 29, 191
Osmotic adjustment, 23, 27–30, 32, 37, 38, 41, 42, 43, 97, 99, 191
Osmotic potential, 3, 5, 6, 7, 27, 29, 32, 37, 39, 55, 65, 70, 94, 95, 96, 98, 191
Osmotic signal, 148
Osmotic stress, 49, 175
Osmotic volume, 29, 50
Oxalate, 97
Oxidation–reduction, 77, 78, 86, 135
Oxidative phosphorylation, 124
Oxygen, 18–21, 24, 52, 57, 77, 98, 100, 110, 113
  in troposphere, 104
Ozone, 98, 101, 177, 181
Ozone layer, 103

Palisade cells, 110
Panicle initiation, 14
Panicum miliaceum, 125
Passive transport, 134
Pathogenesis, 4, 168, 191
Pathogenic organisms, 75, 87
Pathogens, 4, 123, 162, 163, 191
Peanuts, 175
Pearl millet, 161, 167
Pear trees, 174
Peas, 146, 152, 159, 161
Pepper, 97
Peptides, 77
Pericycle, 9, 151
Permanent wilting point (PWP), 7–11, 191
Permeability, *see* Membrane permeability
Permeases, 134, 143
Peroxidases, 52, 78, 164
PGR, *see* Plant growth regulators (PGR)
pH, 72, 83, 85, 89, 139, 143
Phaseic acid, 150, 167, 173
Phaseolus vulgaris, 153, 167
Phase separation, 47, 48, 49, 191
  lateral, 66, 136, 143
Phase transition, 47, 52, 136, 138, 139, 191
Phenolic compounds, 106, 124, 159, 163, 164
Phenol oxidases, 163, 164
Phenylmercuric acetate, 173
Phloem, 74, 77, 96, 100

Phosfon, 175
Phosphates, 78, 79
Phosphatidic acid, 139
Phosphatidylcholine, 138
Phosphatidylglycerol, 139
Phosphatidylserine, 138
3-phosphoglyceric acid (PGA), 16, 17
Phosphohydrolases, 96
Phospholipids, 130, 132–140, 173
Phosphorus, 78, 84, 89, 96, 100, 101, 154, 168, 176
Photoinhibition, 20, 110, 114
Photomorphogenesis, 103, 158, 167, 169
Photons, 104, 112
Photoperiod, 115, 157, 158, 161
Photophosphorylation, 191
Photoreceptors, 105
Photorespiration, 16, 17, 18, 19
Photosensitivity, 113
Photosynthates, 14, 97, 109
Photosynthesis:
  allelochemical effects, 124
  drought effects, 15, 19, 23, 25, 26, 172, 173
  irradiation effects, 103, 106, 110, 112, 114
  nutrient stress effects, 77
  temperature effects, 46, 58, 59, 60
Photosynthetic acclimation, 61, 63, 67
Photosynthetic energy for transport, 134, 135
Photosystem I, 108
Photosystem II, 77, 108
Phototropism, 159
Physical phase transition, 47, 52
Physiogenesis, 4, 191
Physiogens, 4, 191
Phytochrome, 137, 141–144, 157–161, 166, 169
Phytochrome Pr form, 157, 158
Phytochrome Pfr form, 142, 157, 158, 159
Phytoferritin, 79
Phytohormone, *see specific phytohormones*
Phytotoxicants, 121
Pine trees, 85
Pinto beans, 177
Pisum sativum, 174
Plant growth regulators (PGR), 171, 175, 176, 177, 180, 181, 182

Plasmalemma, 12, 48, 129
Plasmolysis, 49, 50, 59, 191
Plasmometric techniques, 29
Plastocyanin, 77
Plastoquinone, 110
Ploidy, 113
Plum, 146
Poikilotherm, 3, 45, 54, 192
Poinsettia, 175
Polarity, 133, 134
Polarizability, 88
Polar translocation, 146
Pole beans, 175
Polyamines, 174, 180
Polyethylene, 172
Polyethylene glycol, 175
Polyphenoloxidases, 77, 163
Polyploids, 113
Polyribosomes, 46
Polysaccharide hydrolases, 163
Polysaccharides, 84
Polyvinyl waxes, 172
Poplar, 161
Porphyrin, 79
Potassium (K), 52, 82, 84, 95, 99, 139, 140, 176
Potassium pump, 52, 140
Potatoes, 45, 54, 65, 86, 92, 154, 177
Predisposers, 122
Pressure gradient, 74
Proline, 20, 24, 26, 41
Proso millet, 125
Proteins, 20, 23, 25, 68, 108, 114, 176
  in acclimation, 46
  in membranes, 47, 50, 52, 59, 63, 64, 66, 129, 130, 132–136, 138, 140, 141, 142
  soluble, 52, 55, 58, 64, 65
  of stress, 179, 180, 181
  structural, 60
  UV effects, 111, 112
Protein synthesis, 13, 21, 63, 64, 65, 67, 111, 124, 179
Protonated membrane systems, 139
Proton gradients, 134
Proton pump, 89
Protons, 89
Protoplast lysis, 46
Pyrimidine bases, 111

Quantum yield, 106

Radiant energy, 14, 16, 103
Radiation, 1, 3, 35
  absorption of, 106, 113
  bright light, 110, 114
  distribution of, 104, 105
  infrared, 106
  ionizing, 112, 115
  light deficit, 36, 103, 109, 114
  solar, 103, 104, 106, 113
  stress resistance, 109, 110, 112, 113
  UV, 104, 106, 111, 112, 114
Radish, 148, 152
Raphanus sativus, 174
Raspberry, 174
Recombinant DNA, 178, 192
Red light (R), 106, 157, 159, 160, 161
Reflection of light, 104, 106, 109, 111, 112
Refraction, 104
Regulators, 46. *See also* Growth regulators (PGR)
Relative growth rate, 125
Relative water content (RWC), 12, 39
Resistance, 45, 192. *See also* Tolerance
  diffusive, 17, 35, 36, 38–41
  to disease, 164
  drought, 3, 10, 24, 25, 27, 32, 33, 37, 39–43, 69, 180
  to frost, 50, 54
  to glyphosate, 178
  to high temperature, 180
  to irradiation, 3, 109, 110, 112, 113
  to low temperature, 50, 64, 65, 172
  to ozone, 98, 177
  and PGR, 171
  protoplasmic, 33
  to salt, 3, 97, 102
  to stress injury, 23, 50
  to water flow, 9, 10, 36, 151
Respiration, 109, 111, 134
  anaerobic, 21
  dark, 17, 107
  energy flow, 77, 135
  photo-, 16, 17, 18, 19
Respiration rate, 57, 59, 61, 64, 107
Rhizosphere, 7, 74, 85, 87, 120, 192
Ribonucleic acid (RNA), 112
Ribosomes, 47

Ribulose bisphosphate carboxylase–
oxygenase, 16, 17, 19, 110
Rice, 85, 97, 148, 159, 164, 166, 175, 179
Root, 21, 24, 39, 43, 91, 93, 102, 178
mycorrhizal, 87, 90, 176
and phytohormones, 146, 148, 151, 153,
154, 159, 163, 166–168
Root absorption, 6, 8, 9, 10, 39, 73, 74, 78,
85, 87, 90, 96, 99
Root exudation, 89, 120, 122, 126, 127, 154
Root growth, 7, 8, 10, 22, 36, 40, 74, 84,
89, 125, 153, 161, 162, 163, 166, 172,
176
Root hairs, 9, 23, 74, 89
Root injury, 22, 84, 118
Root–shoot ratio, 36, 96, 172, 173
Root systems, 36, 40, 118
Rose, 161, 169
Rosette of leaves, 155
Rust disease, 40, 162
Rye, 50, 51, 67, 69, 85, 89

SADH (succinic acid-2,2-
dimethylhydrazide), 173, 176
Salicornia rubra, 95
Salicylaldoxime, 173
Salinity, *see specific Salt entries*
Salmonella typhimurium, 178
Salt excluders, 3, 96
Salt excretion glands, 95
Salt intolerance, 95
Salt nonexcluders, 76
Salt stress, 3, 83, 93, 94, 96, 97, 99, 100
Salt tolerance, 3, 93, 95, 97, 98, 100, 102
Scopoletin, 125
Secondary metabolic compounds, 118, 122,
126
Seeds, radiation sensitivity, 113
Semiarid, 121, 124
Senescence, 113, 120, 148, 162, 163
Serine, 20, 21
Severe deficiency, 72, 77
Severe stress, 2, 11, 12, 17, 20, 27, 36, 42
Shade plants, 106–110, 113
Sieve tubes, 96, 100
Singlet oxygen, 110
Sink, 12, 20, 96
Sitosterol, 131
Snap beans, 175

Sodium, 94, 95, 96, 99, 101
Solanum acaule, 50, 70
Solanum tuberosum, 50, 70
Solar constant, 103, 192
Solar radiation, *see* Radiation
Solid gel, 47
Sorghum, 16, 23, 26, 40, 41, 42, 97
Soybeans, 15, 24, 25, 26, 41, 42, 85, 175,
176, 179, 180
Spartina, 95
Spinach, 175, 177
Stele, 192
Sterols, 52, 131, 132, 136, 137, 139, 142,
143, 144, 177
Stigmella argentipedella, 164
Stomatal aperture, 10, 11, 16, 18, 19, 20,
22, 35, 36, 37, 38, 39, 41, 148, 150,
151
Stomatal conductance, 21, 110, 192
Stomatal guard cells, 50, 151
Stomatal resistance, 35, 38, 192
Stomates, 40
Strawberry, 174
Stress, defined, 1–4, 192
Stress proteins, 179, 180, 181, 193
Stunting, 84, 86, 162, 164
Succinic acid-2,2-dimethylhydrazide
(SADH), 173, 176
Sugar beet, 98, 100
Sugar pump, 52
Sugars, 32, 34, 35, 96, 135
Sulfate ions, 99
Sulfhydryl groups, 52
Sulfur, 90
Sunflower, 15, 22, 24, 25, 148, 154
Sun plants, 106–110
Super cooling, 46, 47, 48, 53, 54
Surfactants, 46
Sycamore, 154
Symplast, 9, 151, 193
Symplococcus spicata, 85
Symport permease, 134, 193

Tea, 85
Temperature, 1, 3, 28, 42, 43, 79
on membranes, 46, 47, 49, 51–55, 59, 61,
132, 135–139, 143, 144
and phytohormones, 63, 64, 66, 69, 70,
152, 153, 154, 157

on roots, 55, 56, 57, 165, 166, 167
Temperature acclimation, *see* Acclimation
   to temperature
Terpenoids, 121, 124
Threshold water potential, 147, 149
Thylakoids, 47, 63, 107, 135
Tillers, 14
Tobacco, 75, 97, 148, 152, 164, 175, 177,
   178
Tobacco mosaic virus, 164
Tolerance, *see also* Resistance
   of air pollutants, 177
   defined, 193
   of dehydration, 27, 28, 38
   of desiccation, 37, 39, 42
   differential, 85
   of drought, 28, 35, 36, 37, 41, 58, 69,
      140, 146, 165, 172, 173, 176, 178, 180
   of flooding, 22
   of freezing, 45, 49, 52, 69
   of frost, 54, 55, 174
   of high temperatures, 39, 46, 56, 58, 59,
      63, 174, 179
   of iron deficiency, 89
   of nutrient toxicity, 85, 86, 87
   of low temperature, 58, 173, 176
   protoplasmic, 37, 42, 43, 54
   of salt, 93, 95, 97, 98, 100, 102, 175,
      176
   of shade, 107, 114
   stress, 2, 3, 179, 180, 181
   and stress proteins, 179, 180
Tomato, 21, 24, 46, 68, 98, 102, 118, 126,
   146, 148, 150, 152, 154, 155, 162, 167,
   173, 175, 176, 178
Tonoplast, 12, 15, 95
Toxicity, 71, 72, 74, 78, 83–87, 90, 91, 92,
   94, 154, 157
Transition metals, 77, 78, 83, 86
Transition temperature, 136–139, 143
Translocation:
   of assimilates, 12, 21, 26, 124
   of nutrients, 20, 71, 73, 74, 78, 79, 82, 86,
      142, 157, 162, 176
   of PGR, 56, 146, 147, 148
   PGR on, 157, 162, 176
   in phloem, 74, 77, 96
   temperature on, 55, 56, 152
   of water, 10, 12

Transmittance of light, 106, 109
Transpiration, 6, 7, 11, 20, 23, 36, 39
   cuticular, 38, 41
   evaporative cooling, 60
   PGR on, 172, 173, 175, 180
   on translocation, 74, 96, 150
Transpiration potential, 10
Transport:
   active, 46, 55, 124, 134, 135, 139, 142,
      143
   of auxin, 146, 151, 156, 159
   of calcium, 168
   of CO_2, 111
   co-transport, 135
   electrons, 77, 110, 111, 135, 141
   facilitated, 134
   ion, 133, 135, 140, 141
   kinetics, 99
   of leucine, 67
   of lipid, 137
   nonspecific, 141, 142
   passive, 134
   permeases, 134
   phloem, 22
   of salt, 102
   of toxin, 126
   of water, 9, 142
Triticale, 89
Trophosphere, 104
Tryptophan, 155
Tulip topple disease, 139
Turgor:
   adjustment, 14, 97
   defined, 193
   full, 28, 29
   loss of, 3, 6, 11, 13, 14, 23, 36, 149,
      150
   maintenance, 27, 28, 32, 37, 43, 95, 96,
      97, 100, 175
   potential, 6
   pressure, 28, 32, 41
   regulation, 29, 139
   zero, 28, 29
Tylosis, 162

Ulmus americana, 22
Ultraviolet radiation, 104, 106, 111, 112,
   114, 172, 180
Uniport permeases, 134, 193

Unsaturated fatty acids, 132, 136, 137, 138, 193

Vapor pressure gradient, 48
Vernalization, 152
Visible light, 104, 106
Vomifoliol, 173

Water:
  activity regulation, 29
  core, apple, 139
  deficit, 38, 39, 146–150
  of hydration, 12, 138, 139
  logged, 146, 148
  on membrane permeability, 133, 136, 140, 142
  potential, 5–15, 21, 23, 24, 26, 28, 29, 30, 32, 35, 37–43, 74, 94–97, 175, 194
  phytohormone effects, 146–152, 161, 165
  saturation deficit, 38, 147
  saver plants, 38
  spender plants, 38
  stress, 4, 36, 37, 38, 43, 145, 146, 147, 152, 163, 165, 166, 167, 168, 178, 194
Weeds, 118, 124, 125, 126
Wheat, 15, 26, 28, 30, 32, 40, 45, 49, 54, 65, 66, 67, 70, 85, 88, 89, 92, 98, 100, 146, 147, 154, 161, 169, 173, 174, 175, 176, 179
White beans, 177
Wilting, 162, 194
Woody species, 22, 46, 54, 69, 87, 99

Xanthophyll, 111
Xanthoxin, 173
X-rays, 112
Xylem, 6, 9, 14, 20, 22, 74, 77, 124, 146, 148, 151, 153, 162, 166

Zeatin riboside, 153, 154
Zero stress, 2, 194
Zinc, 83, 89, 155
Zucchini, 175